04940

Basic Vibration Control

Co-Authors:

T. P. C. Bramer, B. Sc. (Eng), C. Eng., MIEE
G. J. Cole, MCIBS
J. R. Cowell, B. Sc., B. Arch., M. Sc., RIBA, MIOA
A. T. Fry, B. Sc., ARCS, M. Inst. P.
N. A. Grundy, B. Sc., M. Inst. P., MCIBS, MIOA, M. Inst. M.
T. J. B. Smith, B. Sc., Ph. D., C. Eng., MIEE, M. Inst. P., FIOA
J. D. Webb, B. Sc. (Eng), Ph. D., C. Eng., M. I. Mech. E., FIOA
D. R. Winterbottom, B. Sc., MIOA

Sound Research Laboratories Limited
Holbrook Hall, Little Waldingfield, Sudbury, Suffolk, CO10 0TH

First Published 1977
by Sound Research Laboratories Limited
Holbrook Hall, Little Waldingfield,
Sudbury, Suffolk.

© Sound Research Laboratories Limited

ISBN 0 419   114408

Distributed by E. & F.N. Spon Limited
11 New Fetter Lane, London EC4P 4EE.

Distributed in the U.S.A.
by Halsted Press, a Division
of John Wiley and Sons, Inc., New York.

Contents

Chapter	Title	Page
	Introduction	
1	Basic Principles	1
2	Sources of Vibration	11
3	Vibration in Buildings	17
4	Isolator Selection	31
5	Installation	37
6	Plantroom Design and Services Installation	63
7	Special Cases	81
8	Vibration Measurement - Techniques and Instrumentation	91
9	Overall Vibration Control	109
	Notes on the Authors	113
	Terminology	117
	Bibliography	123
	Index	127

Introduction

Vibration control is a subject which concerns architects and engineers in a variety of fields. This book, therefore, sets out to provide a basic grounding in the subject such that the reader may apply much of the information to his own specialist area. In addition to basic vibration theory, an attempt is also made to cover in detail the needs of the architect and building services engineer.

As with all SRL's books, the authors have approached the subject in a practical manner — with the content based on their many years of collective experience. The very important aspects of installation and selection of vibration equipment are discussed at length with many illustrations and charts. After which a chapter dealing with special cases is included to enable engineers to understand the particular isolation requirements of specialist equipment. This section covers factors not usually included in vibration isolator manufacturers information, but nevertheless requires careful attention if a successful installation is to be provided.

The measurement and analysis of vibration has also been covered in a separate chapter. This firstly describes vibratory motions and then goes on to discuss instruments and techniques available for their measurement.

The final sections of the book include an 'aide-memoire' for those concerned with the design of vibration control systems, and a bibliography and list of useful terminology.

Obviously, in any book which attempts to deal with such a wide subject, more information has to be excluded than can be reasonably accommodated. The authors hope that their selection will at least provide a useful introduction to the subject; an introduction both with academic and practical value.

Chapter 1
Basic Principles

The Object of the Exercise

1.   To protect a building or structure from the vibration effects of equipment.

2.   To protect equipment from the vibration effects of the building.

The first case covers the vibration isolation of equipment like fans, pumps, chillers, etc., while the second case usually involves delicate items, such as electron microscopes, ultraprecision machine tools, etc. In both cases there is the additional requirement that the movement of the isolated equipment must be kept down to suitable limits. Fig. 1.

Basic Principles

The basic principle in both of the above cases is the same, the rigid connection between the vibrating system and the protected one is broken. To provide perfect isolation one would have to support the fan or electron microscope on sky hooks. As this is not possible, some form of spring is used instead! In general, the stiffer the spring, the less effective the vibration isolation.

Let us consider the first case in more detail (i. e. a fan which has been mounted on springs with the object of providing vibration isolation for the building). If one tries to push the fan up and down on its isolator at a low <u>frequency</u>, the movement is resisted by the <u>stiffness</u> of the spring mount. The force (F) required to produce a given movement (x) will be given by:

$$F_{spring} = kx$$

where k is the spring stiffness

i.e. if the mounts have stiffness of 100kg per mm, a force of 200kg will be needed to give a deflection of 2mm.

On the other hand, if we try to move the fan up and down at a very high frequency so that it is always stopping and starting, then the main resistance to movement will come from the <u>inertia</u> of the fan. If we move the equipment up and down with an amplitude A in a sinusoidal manner at a frequency of f cycles per second the movement will be given by:

1

$$x = A \sin(2\pi ft)$$

where x is the displacement from the centre position at time t.

Thus the force to overcome the spring stiffness will be:

$$F_{spring} = kx = k A \sin(2\pi ft)$$

The inertia force is equal to the mass times the acceleration. The acceleration, a, for sinusoidal motion (simple harmonic motion) is given by:

$$a = -x (2\pi f)^2$$

Therefore the inertia force is given by:

$$F\ inertia = -\frac{w}{g}(2\pi f)^2 x = -\frac{w}{g}(2\pi f)^2 A \sin(2\pi ft)$$

The negative sign in the equation indicates that the resisting force operates in the opposite direction to the spring force. Whilst the spring force tends to push the fan upwards at the bottom of its stroke back towards its natural position of rest, the inertia tends to make the fan keep on going upwards at the top of its stroke.

The resistance to motion due to inertia increases from zero at zero frequency to very high values at high frequencies. On the other hand, the spring force is independent of frequency. At high frequencies, the inertia force is many times greater than the spring force. Since one force remains constant over the frequency range while the other opposing force increases from zero to a high value, there must be a frequency between two extremes where the two restraining forces cancel out. At this frequency, in theory zero force is required to produce very large movements of the fan (unless friction or some other damping effect restricts the movement). This frequency is known as the natural frequency of the system and if the fan is pulled away from its natural position of rest and released, it will vibrate up and down at that frequency. By combining the two equations, given above, the natural frequency f_o can be expressed as follows:

If the spring force and the inertia force cancel out

$$F\ spring + F\ inertia = 0 \quad for \quad f = f_o$$

or $$k A \sin(2\pi f_o t) - \frac{w}{g}(2\pi f_o)^2 A \sin(2\pi f_o t) = 0$$

$$\text{or} \quad k = \frac{w}{g}(2\pi f_o)^2.$$

$$\text{or} \quad = \quad f_o = \frac{1}{2\pi}\sqrt{\frac{kg}{w}}$$

$$\text{or} \quad f_o = K\sqrt{\frac{1}{d}}$$

where d is the deflection of the mounts under the deadweight of the fan.

Values of K are as follows:

$$f_o = 15.8\sqrt{\frac{1}{d}} \text{ Hz} \quad \text{(d in millimetres)}$$

$$f_o = 3.13\sqrt{\frac{1}{d}} \text{ Hz} \quad \text{(d in inches)}$$

$$f_o = 948\sqrt{\frac{1}{d}} \text{ cycles per minute (d in millimetres)}$$

$$f_o = 188\sqrt{\frac{1}{d}} \text{ cycles per minute (d in inches)}$$

Note that the natural frequency depends on the deflection of the springs under the dead weight of the fan, and on nothing else. This is similar to the case of a clock pendulum where it is only the length that is important.

If the fan was to run at such a speed that an out-of-balance force coincided with the natural frequency, the fan would move up and down by a very large amount. The movement would increase either until it was limited by damping, non-linear effects, or until something broke. This condition is known as resonance.

If we consider our spring-mounted fan starting up slowly from rest and running at progressively higher and higher speeds, we can examine the effectiveness of the springs as vibration isolators. The force transmitted downwards by the spring will be proportional to the amplitude of the movement, i.e. if the movement is large, the transmitted force will be large. At low frequencies the fan will be "stiffness controlled" and the movement will correspond with the stiffness of the springs. The transmitted force will therefore be equal to the out-of-balance force, i.e. no isolation. As the frequency of rotation increases the inertia force will start to cancel out the stiffness force - the amplitude will increase and the transmitted force will therefore also increase. This will continue until resonance is reached, at which point the transmitted force is very much higher than the out-of-balance force. As the disturbing frequency (f) is increased further, the inertia force becomes greater than the stiffness force and the movement and transmitted force reduce. At this stage the system is said to be "mass controlled", in that it is the mass that controls the movement of the fan. It is not until this state is reached that useful vibration isolation occurs. No isolation whatsoever occurs at frequencies below 1.4 times the resonant frequency. Up to this point vibration isolators cause amplification. Therefore, for vibration isolation to be effective, the natural frequency of the isolated system must be much less than the disturbing frequency. Figs. 2 and 3. For ideal systems the degree of isolation depends only on the static deflection and on the forcing frequency. Fig. 4.

The Three Controlling Factors

The three controlling factors in a basic vibration isolation system are the stiffness of the springs, the mass of the suspended equipment, and the damping of the system. Their effects can be summed up as follows:

1.   Stiffness

The springs provide the isolation; the stiffer the springs, the less effective the vibration isolation. Rather surprisingly (unless you look at the equations), the springs have little effect on the amplitude of motion for a properly operating vibration isolation system.

2.   Mass

The mass keeps the suspended system still, the heavier the suspended mass, the smaller the movement for a given disturbing force. Since the heavier mass requires stronger springs to support it, for a given resonant frequency, increasing the mass by means of an inertia block does not reduce the transmitted force if the static deflection and resonant frequency are kept constant. It does, however, reduce the movement of the suspended system.

3.   Damping

Damping has three effects

(a)   it reduces the effect of resonance

(b)   it reduces the amplitude of motion of the suspended system at high frequencies

(c)   but on the debit side it tends to increase the transmitted force by effectively increasing the spring stiffness by providing an additional connection short circuiting the spring

Transmissibility and Isolation

The effectiveness of isolation can be expressed in terms of the proportion of the initial force which is transmitted. This is known as the transmissibility

$$\text{Transmissibility } T = \frac{F_t}{F_o}$$

where F_t is the transmitted force and F_o is the disturbing force. Vibration isolation occurs when $T < 1$

This is sometimes expressed in terms of efficiency as given below:

$$\text{Efficiency } \eta = \frac{F_o - F_t}{F_o} = 1 - T$$

(Note: efficiency and transmissibility are usually expressed as percentages).

Transmissibility is the better concept in that it gives a better picture of the degree of goodness of a system. This can be seen if the equivalent transmissibilities and efficiencies are set out side by side as below:

Disturbing Force (F_o) (Kg)	Transmitted Force (F_t) (Kg)	Transmissibility (T) %	Efficiency %
10	10	100	0
10	5	50	50
10	2	20	80
10	1	10	90
10	0.1	1	99
10	0.01	0.1	99.9

An isolation efficiency of 99% does not sound much better than an efficiency of 90%, but one sees from the transmissibility that it corresponds to a reduction by a factor of 10 in the transmitted force.

For low damping systems: e.g. steel springs

$$\text{Transmissibility } (T) = \frac{1}{\left(\dfrac{f}{f_o}\right)^2 - 1}$$

With viscous damping

$$\text{Transmissibility (T)} = \sqrt{\frac{1 + 4D^2 \left(\frac{f}{fo}\right)^2}{\left(1 - \left(\frac{f}{fo}\right)^2\right)^2 + 4D^2 \left(\frac{f}{fo}\right)^2}}$$

D is the "damping ratio" and compares the actual damping present C to that required for critical or "dead beat" damping Co; $\left(D = \dfrac{C}{Co}\right)$

Typical values of D are:

Material	D
Steel	0.005
Natural Rubber. 70 Shore	0.04
Natural Rubber. 40 Shore	0.02

These equations are plotted in Figs. 2 and 3.

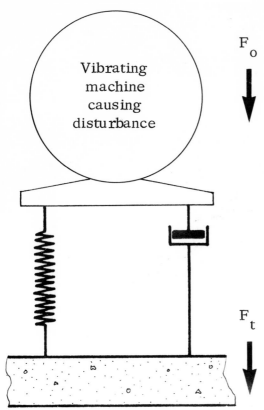

Case 1

Protecting the structure
from the machine

Case 2

Protecting equipment
from structural vibration

x_o = displacement of foundations

x_t = resultant machine vibration amplitude

F_o = vibratory force acting on the machine

F_t = force transmitted to the supporting structure through the isolator

Fig 1

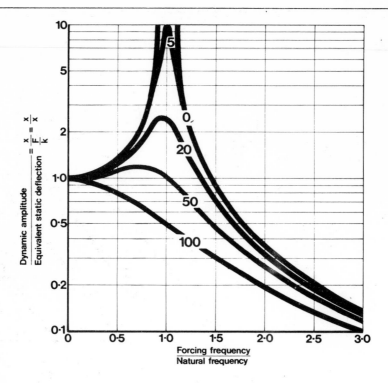

Fig 2   Amplitude of vibration for a viscously damped system

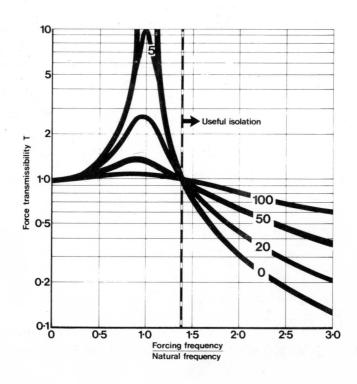

Fig 3   Transmissibility for a viscously damped system

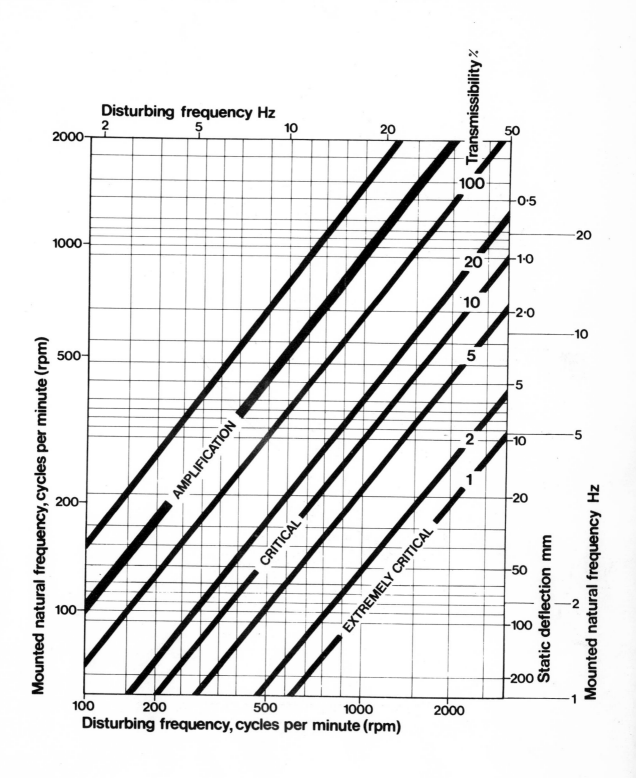

Fig 4

Chapter 2
Sources of Vibration

We must look at two main areas -

1. Causes of vibration in plant and machinery.
2. Types of machinery producing vibration and their situation.

1. Causes of Vibration

There are two main reasons for vibration in machinery both involving out-of-balance forces - directly rotating machinery which is statically or dynamically out-of-balance, such as motors and fans, etc., and reciprocating machinery which is inherently unbalanced, such as internal combustion engines and compressors, etc. A third source is impact and shock from punch presses, etc.

All the above cause vibration as an unwanted by-product of their normal operation but there is a further category of machines which produce vibration deliberately as part of their operation. This includes vibrating screens, conveyors and bowl feeders for small parts in assembly.

a) Rotating Machinery

A simple rotor consisting of a thin circular disc is said to be statically unbalanced if its axis of rotation does not pass through the centre of gravity (C of G) of the rotor (Fig 1a). This situation causes an out-of-balance force which rotates at rotor speed and causes a reaction on the bearings. It can be detected by a simple static test and corrected by the addition of an appropriate balancing mass on the opposite side of the axis (or by removing mass from the 'heavy' side).

If the rotor is not a simple thin disc but has width and the out-of-balance mass consists of two equal parts which are placed as shown in Fig 1b, the rotor will be statically balanced. If, however, it is rotated there will be a force tending to cause the shaft to rotate in a 'skew' manner and produce alternating forces on the bearings. In practice, both static and dynamic unbalance are found in rotating machines and they can be corrected by the addition of a maximum of two single balance weights in appropriate positions.

In the case of a fan, for instance, there is an additional source of unbalance if the rotor is not aerodynamically symmetrical. Differing aerodynamic forces on dissimilar blades can cause static and dynamic type out-of-balance problems.
In all cases of rotating out-of-balance, the lowest frequency of vibration produced is at shaft rotation speed, i.e. $f=\dfrac{rpm}{60}$ Hz. Higher harmonics are also produced

in most cases.

Another problem which must be considered is the critical 'whirling' speed of a rotor. Due to flexibility of the shaft, and originating perhaps from a normally insignificant out-of-balance mass, the shaft will start to bow at a particular speed. This situation is unstable and is known as the critical whirling speed (see Fig 2). Normally this can be predicted and is designed to be well above any normal running speed. However, the simple theory depends on the bearings being perfectly rigid. In practice they are usually fixed to a frame or baseplate which allows small movements. This non-rigidity of bearings can sometimes cause a critical whirling speed to be within the normal range of the machines' operating speeds. A small amount of 'whirling' can lead to excessive vibration and the possibility of a permanently bent shaft. Troubles due to 'whirling' can sometimes occur when the speed of a centrifugal fan is increased above that recommended by the manufacturer in an effort to increase output.

Fans and motors are normally carefully balanced in manufacture or on site but imbalance can occur from a number of causes such as uneven build-up of dirt on fan blades or uneven erosion of fan blades in abrasive air flows.

It must be pointed out that the resultant out-of-balance forces and couples in practical machines can occur in a number of modes which must be considered. In most cases, isolation of the simple modes with practical isolation devices will automatically cope with the other modes, but there are cases where these must be considered.

b) Reciprocating Machinery

The problem here is much more complicated than simple rotating machinery and could form the basis of a complete book! However, the problem can be simplified by stating that the addition of inherently unbalanced reciprocating masses to the simple rotating shaft in an engine or compressor leads to a complexity of forces and couples at various frequencies.

In practical terms with multi-cylinder engines and compressors, the results simplify into inertia forces in the three mutually perpendicular axes and couples about these same axes. This is shown in Fig 3. In practice, five of the six effects are important - the inertia force in the longitudinal direction is usually insignificant. Because of the varying angle that the connecting rod makes with the cylinder axis, the reciprocating forces are not purely sinusoidal and harmonics of forces and couples are produced. This leads to primary forces and couples which are periodic at crankshaft rotational speed with higher order harmonics which are called secondary and tertiary, etc., orders. In practice, the only important ones to consider are the primary and secondary effects at engine speed and twice engine speed.

Differing numbers of cylinder configurations and crank arrangements lead to the situation that some engine and compressor types have inherent balance for certain forces and couples. Some of the inherent unbalance effects can be fully or partially balanced by various means and this is always done to the practical limit to reduce harmful and inconvenient vibration. Some of the common engine and compressor configurations are listed in Fig 4 with the resultant balance situation. In some cases, different crank and cylinder arrangements from the usual ones give different results.

The out-of-balance forces above produce vibration effects but there is another source of vibration (and airborne noise). This is caused by the firing explosions in an engine or the gas pressure pulsations in a compressor. This could occur at a lower frequency than shaft rotational speed in a simple machine with one or two cylinders. However, in common multi-cylinder machines it occurs at shaft or higher frequency. For example, in a normal four cylinder four stroke engine it occurs at twice engine speed.

c) Impact and Shock

The above reasons for vibration have been periodic, generated by a rotating source. A third source is from impact in machines such as presses and guillotines, etc. This is a non-periodic or transient effect caused by sudden forces and their reactions. By appropriate analysis it can be resolved into components at various frequencies and dealt with.

2. Types of Machinery and Situations

It must be reasonably obvious by now which types of machinery and plant are sources of vibration and why. A final variety must be dealt with - deliberate vibration for a purpose. There are three main methods of achieving vibratory action on a conveyor, etc:
(i) Direct mechanical out-of-balance vibrator consisting of an eccentric mass rotated by a motor. This principle is normally used for the largest requirements.
(ii) Electro-magnetic vibrator consisting of an armature driven by an alternating electric current - usually the mains at 50Hz. Used in medium applications.
(iii) Air operated vibrator consisting of a mass vibrated by an airstream. Used in small bolt-on devices for high frequencies.

All the above devices should be correctly isolated from the building structure and therefore cause no vibration problem. Normally, however, the isolation is not perfect - particularly if the support is steelwork and not rigid - and they act as vibration sources. In practice, it is usually the first type and sometimes the second that create problems.

Although it will be dealt with elsewhere, it is worth pointing out that the position of a vibrating source in a building is of great importance. For example, a large fan may be perfectly acceptable in a basement plantroom with minimal isolation but a severe problem mounted on steelwork at high level in a building. More will be said of this in the next chapter.

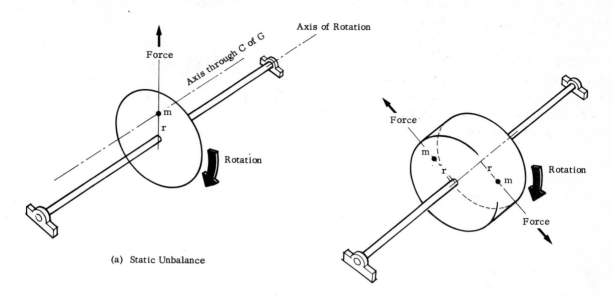

(a) Static Unbalance

(b) Dynamic Unbalance (static balance)

Fig 1

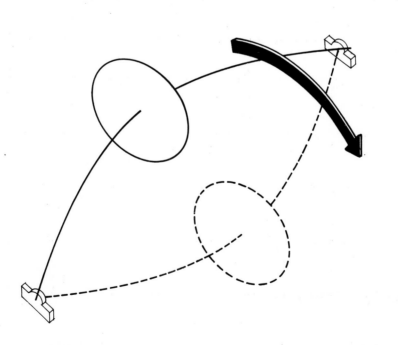

Fig 2   Whirling effect

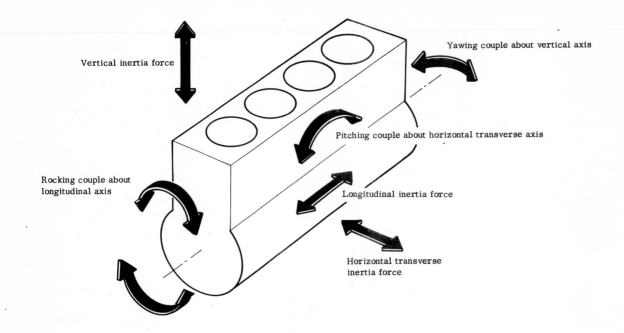

Yawing couple about vertical axis

Vertical inertia force

Pitching couple about horizontal transverse axis

Rocking couple about
longitudinal axis

Longitudinal inertia force

Horizontal transverse
inertia force

Fig 3

CYLINDERS	INERTIA FORCES		COUPLES	
	Primary	Secondary	Primary	Secondary
4 in line	Balanced	Unbalanced	Balanced	Unbalanced
6 in line	Balanced	Balanced	Balanced	Balanced
8 in line	Balanced	Balanced	Balanced	Balanced
8 veeform	Balanced	Balanced	Unbalanced	Balanced

Fig 4   Table showing balance situation for common multicylinder
engine/compressors

Chapter 3
Vibration in Buildings

Introduction

Building technology has developed in the direction of lighter and less rigid construction forms and mechanical plant has become more extensive, more varied and often more powerful. As a result, building construction has become an increasing influence on problems of vibration isolation.

When we examine some of our current buildings, it is clear that - from the point of view of response to vibration - we are dealing with a very complex system. Buildings incorporate an arrangement of components with widely varying properties, e.g. density, moduli or elasticity, span, load distribution, fixings, geometry, stiffness and damping.

The techniques which apply to calculations for vibration control in real buildings become extremely involved and there is still a great deal of statistical research required in this field to develop reliable design procedures. However, experience has provided a number of guidelines for practical application. Provided the basic principles of structural vibration are understood, these guidelines can usually be applied successfully to specific cases.

The aim of this chapter is to provide an understanding of the contribution made by the building structure in mechanical services vibration control problems and to develop practical means of accounting for this.

External Vibration Sources

Our primary concern is with vibration sources associated with the mechanical services system (these have been discussed more fully in Chapter 2). However, it is useful to refer briefly to the presence of other external sources which can also contribute to vibration of the building structure (see Fig 1).

For example, pile driving or underground railways often cause building vibration problems, particularly where the resonant frequency of the ground itself is excited and efficient coupling with the building occurs (see Fig 2). A number of complete buildings have been set on purpose-designed mounts; railway lines have been set on resilient bases and new piling techniques have been developed to counteract the potential problems. Wind sway frequencies are associated with slow fluctuations in wind speed, which, in the context of urban geometry and tall buildings, can set up substantial vibration in a sway mode generally in the range 0-2Hz (Ref. 8). Other internal sources such as dance floors may also require consideration.

Transmission Through The Structure

The propagation of vibration through structures occurs in four basic wave forms which can be combined together to produce complicated movements of building components (see Fig 3). These are:

Longitudinal Waves
Transverse Waves
Torsional Waves
Flexural Waves

Longitudinal waves are derived from successive compression and separation along the longitudinal axis of the member concerned. Transverse waves are similar but operating on axes across the member. Torsional waves involve twist, shear and rotational forces. Flexural waves refer to 'rippling' of the surface of the components - in fact, this is the easiest way for most materials to move and therefore most important for building vibration.

In order to cover the effect of time on vibration processes, we must also distinguish between:

Travelling Waves
Maintained Waves

These two forms are self-explanatory. Instantaneous or short term sources, e.g. piling or underground rail movements will induce travelling waves which are not reinforced by subsequent energy input - therefore a state of equilibrium of building movement is not reached. Maintained waves are the type more likely to concern us in that the plant is usually running for some period of time. Account must of course be taken of the on/off, run-up and run-down cycles of plant.

Attenuation Through The Structure

The attenuation through the structure will be derived from two primary components:

a) Natural Attenuation
b) Resonance Features

a) Natural Attenuation

In general, we find that natural high frequency attenuation is greater than low frequency attenuation within the bulk of most building materials taking typical, relatively large, geometrical forms. High frequency attenuation occurs at discontinuities and junctions. Fig 4 shows an example of the reduction in vibration level with distance along a typical ribbed concrete floor. It must however be stressed that the variability of such data is substantial and accuracy is limited, partly due to the variability of workmanship on site. Natural attenuation is derived from:

Internal Damping
Reflection at irregularities/Poor impedance matching
Spreading of the energy

b) Resonance

The resonance of building structures is an all important item for identity in proper vibration control. Fig 5 shows how components for a building may be induced to resonance and thereby by-pass the natural attenuation. Individual components will have resonances of their own, but by virtue of the compound connections between the components further numerous resonances are produced. Fig 6 sets out the various modes in which resonance occurs for a horizontal member under a number of end fixing conditions. On the basis of theoretical formulae relating to primary building components (eg Refs. 3 and 6) it is possible to analyse building forms using matrices and a computer. While this is most useful in certain circumstances, there are likely to be limitations to the accuracy of results when related to real site conditions, since workmanship can have a major effect on building response.

It is clear from Fig 2 that the typical resonant frequencies of building components overlap with the frequency of operation of slower speed plant. In practice, it is usually more realistic to make an assessment of the range of expected resonance. In the case of machinery mounted on a floor structure, we can obtain a closer estimate of the principal resonant frequency of the slab by reference to the following formula:

$$\text{Resonant frequency} = 15.75 \sqrt{\frac{1}{\text{Deflection (mm)}}} \text{ Hz}$$

This relationship can be used to estimate the probable natural resonant frequency of a floor for the most common systems of construction, by assessing the probable floor deflection and substituting in the equation.

Table 1 offers the probable natural frequency for concrete floor construction for differing span between supports (Ref. 5).

But changes in loading or provision of intermediate supports must be accounted for. The positioning of plant at the end of a cantilever or asymmetrically on a simple slab offers a different condition of deflection, and thereby resonance. As the plant is moved from the simple centre span position, closer to columns, the reduced deflection results in an increase of resonant frequency. The complex sections of some typical slabs often render computation of resonances for asymmetric loading difficult. In practical cases, the structural engineers should be able to provide adequate information. It is worth keeping in mind that vibration isolation for centre span cases is likely to be on the safe side when used for floor loading positions giving higher resonance. Waller (Ref. 8) makes the point that where structural resonance is found in practice to be badly located on the frequency scale, conversion of non-structural partioning to a wholly or partly load bearing prop somewhere near mid span may well effect an adequate adjustment of resonant frequency.

Damping

The amount of amplification occurring at resonance is called the Q factor of the construction. Typical values for Q are (Ref. 4):

Bolted steel structure	20 - 60
Welded steel structure	30 - 100
Reinforced concrete structure	12 - 25

Where pre-stressing and post stressing is employed, not only do we find that spans are often very long and therefore the resonant frequency very low, but also the Q factor is very high due to the constant compression of the concrete and less absorption of energy at shrinkage cracks.

Damping occurs in two forms - internal damping and radiation losses.

Internal damping is related to energy conversion through friction inside the construction. A common internal damping method involves creation of lateral shear forces by using laminar construction. Attempts have also been made to introduce damping materials into concrete mixes.

It is worth noting that with travelling waves the time required to reach full resonance may be longer than the period of excitation. For this reason, it is not always good design to add damping to cope with travelling waves (Ref. 4).

Principles of Mounting

In order to achieve acceptable vibration and sound levels in buildings, we must assess both natural attenuation and resonance. Experience with natural attenuation has provided guidance on the required percentage isolation for differing plant mounted on varying structures, taking account of the location of critical areas relative to the source (see Chapter 4). Therefore, we are normally concerned primarily to avoid resonance.

Fig 7 illustrates the effective role of the support structure as a mass/spring. Table 2 sets out the relationship determining the two resonant frequencies of such a compound system (Ref. 9). These two resonant frequencies will tend to come close together if the lower spring is substantially stiffer than the upper mounting. If the reverse occurs, the resonant frequencies will move apart and may cause compound mounting problems which lead to difficulties (see Fig 8). The design procedure is therefore to keep resonant frequencies close together and well away from the forcing frequency. Usually the compound resonances are close to the slab resonances noted in Table 1 and successful design can be related to ensuring that the static deflection on the mount is at least six times that of the floor. With longer spans this often involves considerable mount deflection which can lead to stability problems. It can sometimes be the better of two evils to fix such plant direct to the structure rather than risk resonance with a smaller static deflection.

Re-radiation of Vibrational Energy

A number of research workers have investigated the relationship between the vibration of room boundaries and the resultant room sound pressure levels. The complex movement of these surfaces renders the overall relationship difficult to assess. However, the following is found to work quite well in practice (Ref. 7):

$$\text{SPL (re } 2 \times 10^{-5}\text{N/m}^2) = V + 10 \log_{10}S + 10 \log_{10}r - 10 \log_{10} \left(\frac{A}{4}\right)$$

where V = velocity level of boundaries (dB re 5×10^{-8}m/sec)
S = area of boundaries (m^2)
r = radiation factor
A = room absorption (m^2)

The radiation factor is of the order of 10^{-1} at frequencies below the critical frequency (for concrete and brickwork - below approximately 70Hz). Above the critical frequency r can be taken as 1. A can be derived from reverberation time and room volume.

It is important to bear in mind the capacity of the structure to amplify vibration fed in to it. Fig 9 illustrates the common case of amplification due to rigid pipe fittings which can be alleviated by resilient clamping.

Conclusions

The interaction of the structure and plant in vibration control is increasing due to design trends. Unfortunately the technology of building vibration control still requires a lot of development and is likely to suffer from the effects of variable workmanship. However, it is most important for mechanical services engineers to establish a basic understanding of building vibration as a context for the design maxims noted in this chapter and the discussions in other chapters.

Span between supports (metres)	Probable floor natural frequency (Hz)
3	12
6	9
9	7
12	6
18	5

Table 1   Probable natural resonant frequencies for concrete floor construction

$$f_{res} = \sqrt{\tfrac{1}{2}\left(f_2{}^2 + f_1{}^2\right) \stackrel{+}{_-} \tfrac{1}{2}\sqrt{\left(f_2{}^2 - f_1{}^2\right)^2 + 4f_1{}^2 \, f_3{}^2}}$$

where $f_1 = 15.75 \sqrt{\dfrac{K_1}{M_1}}$ Hz

$f_2 = 15.75 \sqrt{\dfrac{K_1 + K_2}{M_2}}$ Hz

$f_3 = 15.75 \sqrt{\dfrac{K_1}{M_2}}$ Hz

M_1 = weight of plant (kg)

M_2 effective weight of supporting floor (kg)

K_1 = stiffness of machine mounting (kg/mm)

K_2 effective stiffness of supporting floor

f frequency (Hz)

Table 2   Resonant frequencies of a coupled system

References

Crede, C.   (1957)   'Principles of Vibration Control' - Handbook of Noise Control, Ch. 12. (Editor C. M. Harris) McGraw-Hill

Crockett, J. M. A. and Hammond, R.E.R.   (1947)   'Reduction of Ground Vibrations into Structures' - paper given to Institution of Civil Engineers, 11th March. Structural Paper No. 18

Cremer, L.   (1953)   'Calculations of Sound Propagation in Structures' Acustica Vol. 3, 1953

Grootenhuis and Allaway   (1971)   'Noise and Vibration Nuisance Inside Buildings' SEE symposium Olympia 21st June

Rector, J. T.   (1970)   'Noise and Vibration in Multi-Storey Buildings' The Construction Specifier, October

Stokey, W. F.   (1961)   'Vibration of System having Distributed Mass and Elasticity' - Ch. 7 of Shock and Vibration Handbook (Editors C. M. Harris and C. E. Crede) McGraw-Hill

Tukker, J.C.   (1972)   'Application of a Measuring Method for the Dynamical Behaviour of Building Structures' Applied Acoustics Vol 5, no 4, pages 245-264.

Waller, R.A.   (1969)   'Building on Springs' - International Series of Monographs in Civil Engineering, vol 2 - Pergamon Press

Webb, J.D.   (1971)   'Vibration of Coupled System' - SRL Tech. Note No 2

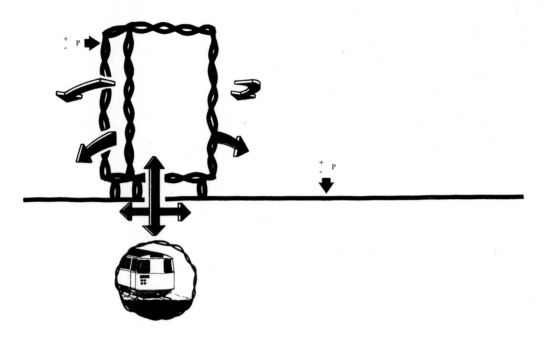

Fig 1   Vibrational movement due to external sources

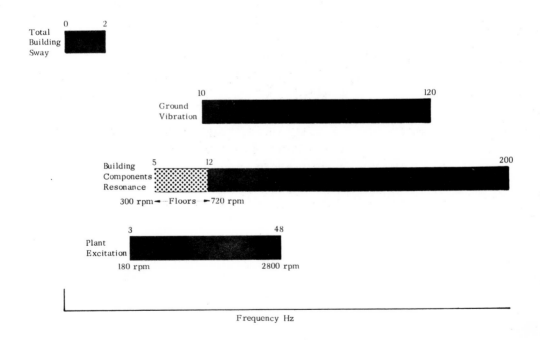

Fig 2   Comparison of natural frequencies of wind sway, ground and
building component movement with plant running speed

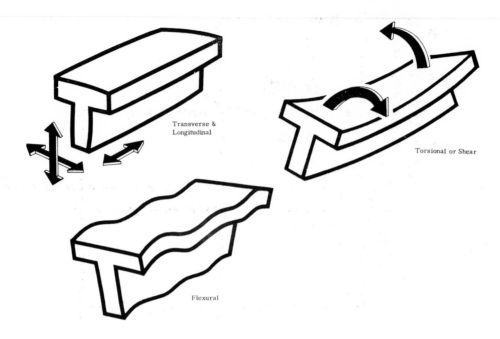

Fig 3   Primary forms of propagation in building structures

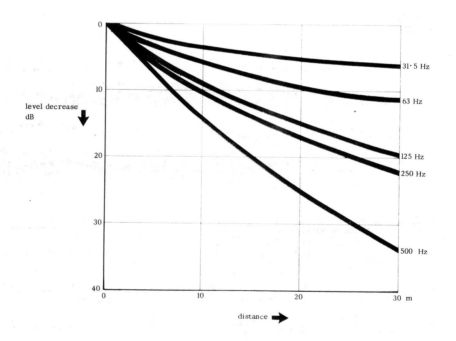

Fig 4   Typical relationship between attenuation rate for different
frequencies along a ribbed concrete floor

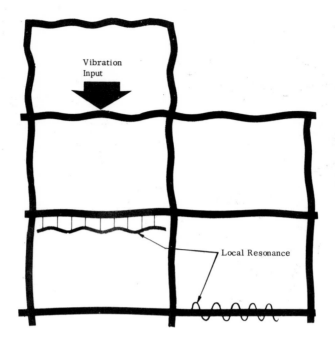

Fig 5   Local resonance

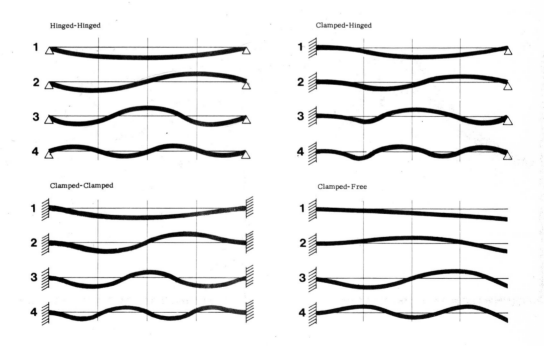

Fig 6   The effect of end fixity on modes of vibration

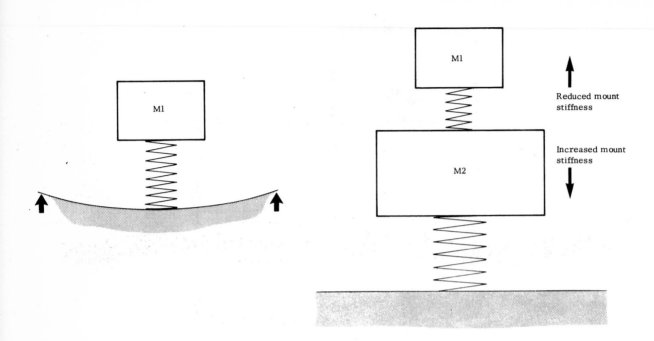

Fig 7   Comparison of structural support with a compound spring system

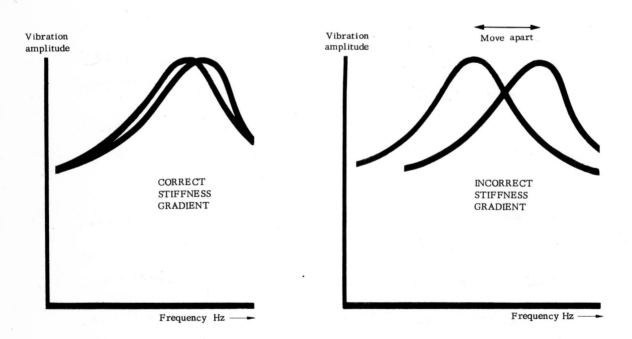

Fig 8   Relative positions of two primary resonant frequencies of a compound system

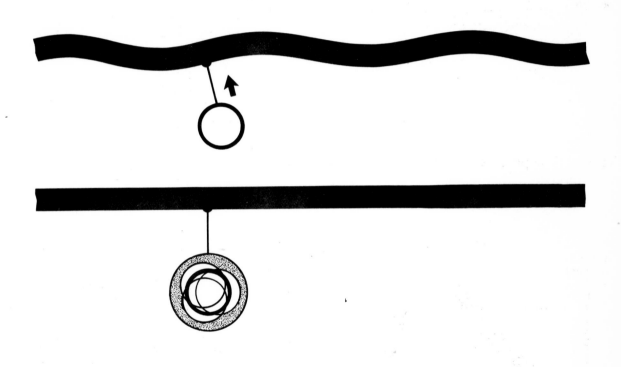

Fig 9   Isolation of pipe vibration from the structure

Chapter 4
Isolator Selection

It should now be clear why isolation of vibrational forces is necessary and the next step is to select a suitable isolator. Firstly it is an advantage if a review of the types of isolators that are commonly used be made.

Perhaps one of the most common means of providing isolation is with some form of padding or matting material. The materials used in this type of isolator range from cork to rubber or neoprene.

The second type of mount usually used in the isolation of mechanical services plant again uses rubber but in this case is used in a form that is commonly called a rubber-in-shear isolator. Here it is usually necessary to physically fix the mounted machine to the isolators and often also the isolators to the supporting structure.

The last kind of common isolator to be generally used in heating and ventilating design is steel springs. These are similar to the previous type but, as the name suggests, springs are used instead of rubber-in-shear.

The first decision to take is to decide how much of the vibrating force can be transmitted to the structure. In the past, buildings have comprised fairly massive structures for the support of the machines to be isolated. General guidelines were evolved indicating the recommended isolation efficiency (and transmissibility) for the machines (Fig 1).

Using the chart shown in Fig 2, with the recommended isolation efficiency, the required static deflection could be determined, knowing the disturbing frequency which was usually taken as the running speed.

Example

What static deflection is required from isolators to support a floor mounted axial flow fan (5HP) running at 1400rpm and located over a critical area?

From Fig 1 the isolation efficiency required is 90% (transmissibility 10%).

Moving to Fig 2, the static deflection required from the isolators is found to be 5mm (0.2in).

This procedure thus enabled suitable vibration isolators to be selected. The range of deflection for the isolators described earlier is approximately as follows:

Isolation mats or pads	up to 5mm (0.2in)
Rubber-in-shear isolators	up to 15mm (0.6in)
Steel spring isolators	up to 175mm (7in)

Many similar selection guides to that shown in Fig 1 are published and these are mainly based on experience with similar equipment and building structure. Most manufacturers offering this type of information will assume responsibility for proper vibration isolation where they supply the isolators required.

There has however been a drastic increase in the cost of building materials and labour over the last decade. Property companies and architects are therefore now turning to new building materials and new construction methods, to speed building and compensate for spiralling costs. Similarly, as building costs increase, the cost of the lettable space that they contain also increases, and less space is being put aside for plant. This has increased the tendency for plant to be located at roof level.

The result is a lighter structure, with the majority of plant located over critical areas. With a requirement for lighter and simpler structures, the spans between supports are likely to increase quite drastically. Hence, as shown in the table in Fig 3, the allowable floor deflection, together with the probable floor deflection that will occur, has also increased. This means that many floor systems could be in resonance with certain plants, especially centrifugal fans, and in these circumstances the selection method referred to previously is no longer adequate. Small disturbing forces in resonance with the structure can cause large movements. It is therefore apparent that vibration isolators must be selected not only to provide deflections sufficient to isolate the disturbing force, but also to have additional deflection to compensate for the increasing deflections to be found in the floors. In some situations isolators with very large deflections are essential to avoid the situation where the floor is more flexible than the isolators, in which case the floor may move up and down instead of the isolated plant!

In specifying isolator performance, a trend is therefore rapidly increasing towards the use of deflection rather than ideal isolation efficiency. The reason being that ideal isolation efficiency takes no account of the deflection of the floor. In order to allow for building structures becoming lighter, a modified table can be drawn up showing static deflection requirements for differing plant on differing floor spans. This is shown in Fig 4.

In the previous example, it can be seen that deflection of 25mm (1in) is called for, which would more than likely result in the use of steel springs rather than rubber-in-shear isolators, and this is the general tendency as floor spans are becoming greater.

Although the cost of steel springs is higher than rubber-in-shear and other isolators having a lower deflection, the increase in cost taken over the project as a whole is minimal. One advantage with springs is that they will not wear out. Organic isolation materials such as cork, rubber, etc., may deteriorate with time and will not ensure continued performance as wear causes increased vibration levels. A further advantage with the selection method, using deflection rather than isolation efficiency, is that deflection can be readily checked by site measurement. It is an almost impossible task to verify the isolation efficiency.

Basically, the more deflection provided by a loaded isolator, the lower its natural frequency and the greater the isolation efficiency. For example, when a car is lightly loaded, the springs will deflect a small amount and there is little isolation from road shock and vibration. When the car is loaded, the springs are fully deflected, resulting in a greater riding comfort.

Properly selected steel springs provide highly efficient isolation of low frequency vibrations. High frequency vibrations such as electrical hum and fan impeller blade impulses can be transmitted through the coils of springs and into the floor structure unless a noise isolation pad is used in series with the springs.

Whilst stable coil springs are linear isolators, spring isolators with built in isolation pads or snubbers, rubber-in-shear and other rubber mounts do not have uniform deflection with load. Often the theoretical isolation efficiency specified is based on the deflection of non-linear isolators. The results are often unsatisfactory and expensive to correct.

Inertia Blocks

One method of mounting equipment in critical locations makes use of concrete inertia bases on which the mechanical equipment is mounted.

The isolators are mounted between the inertia base and the supporting structure and the centre of gravity is effectively lowered. This has the effect of reducing the motion of the mounted equipment when starting up as well as during normal operation. Inertia bases are useful in critical areas on plant having large out-of-balance forces, and which are non-continuous in operation, such as refrigeration compressors.

In summary, the majority of plant vibration problems can be avoided providing consideration is given to:

1. The disturbing frequency of the equipment
2. The natural frequency of the isolators which are to support the equipment
3. The natural frequency of the structure which supports the isolators

This will ensure that resonant conditions are avoided and the desired degree of isolation is provided.

Critical Areas

Critical Areas	Transmissibility	Isolation Efficiency
Centrifugal Compressors	0.5%	99.5%
Centrifugal Fans — greater than 25 HP Reciprocating Compressors — greater than 50 HP Pumps — greater than 5 HP	2%	98%
Axial Flow Fans — greater than 50 HP Centrifugal Fans — 5 to 25 HP Reciprocating Compressors — 10 to 50 HP Pumps — 3 to 5 HP Unit Air Conditioners — Supported Fan Coil Units — Supported	4%	96%
Axial Flow Fans — 10 to 50 HP Centrifugal Fans — up to 5 HP Reciprocating Compressors — up to 10 HP Pumps — up to 3 HP Air Handling Units	6%	94%
Axial Flow Fans — up to 10 HP Unit Air Conditioners — Hung Fan Coil Units — Hung Pipes — Hung	10%	90%
Gas fired boilers (more than 100,000 BThU, 25 kw)		7 to 12 Hz
Oil fired boilers (more than 60,000 BThU, 15 kw)		4 to 7 Hz

Less Critical Areas

Less Critical Areas	Transmissibility	Isolation Efficiency
Centrifugal Compressors	6%	94%
Centrifugal Fans — greater than 25 HP Reciprocating Compressors — greater than 50 HP Pumps — greater than 5 HP Unit Air Conditioners — Supported Fan Coil Units — Supported	10%	90%
Axial Flow Fans — greater than 50 HP Centrifugal Fans — 5 to 25 HP Reciprocating Compressors — 10 to 50 HP Pumps — 3 to 5 HP Air Handling Units Unit Air Conditioners — Hung Fan Coil Units — Hung	20%	80%
Axial Flow Fans — 10 to 50 HP	25%	75%
Axial Flow Fans — up to 10 HP Centrifugal Fans — up to 10 HP Reciprocating Compressors — up to 10 HP Pumps — up to 3 HP Pipes — Hung	30%	70%
Gas fired boilers (more than 100,000 BThU, 25 kw)		12 to 20 Hz
Oil fired boilers (more than 60,000 BThU, 15 kw)		12 to 20 Hz

Fig 1   Recommended isolation efficiencies (concrete floor slab)

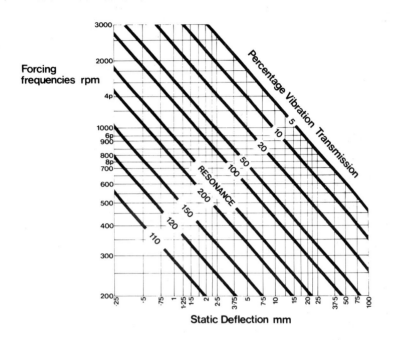

Fig 2   Relationship between percentage transmission, forcing frequency
and static deflection

Span Between Supports	Allowable Floor Deflection (1/300th of Span)	Probable Deflection (20% of Allow)	Probable Floor Natural Frequency	Probable Floor Natural Frequency
10 ft.	0.33 in.	0.066 in.	720 CPM	12.0 CPS
20 ft.	0.67 in.	0.134 in.	560 CPM	9.3 CPS
30 ft.	1.00 in.	0.200 in.	400 CPM	6.7 CPS
40 ft.	1.33 in.	0.266 in.	360 CPM	6.0 CPS
60 ft.	2.00 in.	0.400 in.	300 CPM	5.0 CPS
3.0 m	8.3 mm	1.66 mm	720 CPM	12.0 CPS
6.0 m	17.0 mm	3.4 mm	560 CPM	9.3 CPS
9.0 m	25.0 mm	5.0 mm	400 CPM	6.7 CPS
12.0 m	33.3 mm	6.66 mm	360 CPM	6.0 CPS
18.0 m	50.0 mm	10.0 mm	300 CPM	5.0 CPS

Fig 3

Equipment Type	Equipment Location				
	On Grade	On 6m Floor Span	On 9m Floor Span	On 12m Floor Span	On 15m Floor Span
	Min. Defl. mm	Min. Defl. mm	Min. Defl. mm	Min. Defl. mm	Min. Defl. mm
Refrigeration Machines					
Absorption	6	25	25	45	45
Packaged Hermetic	10	25	45	45	65
Open Centrifugal	10	25	45	45	90
Reciprocating:					
500-750 RPM	25	45	45	65	90
751 RPM & Over	25	25	45	65	65
Compressors					
Air or Refrigeration					
500-750 RPM	25	45	65	65	90
751 RPM & Over	25	25	45	65	65
Boilers or Steam Generators					
All	6	25	25	45	65
Pumps					
Close Coupled thru 5 HP	10	25	25	25	25
Close Coupled 7.5 HP & Over	25	25	25	45	45
Base Mounted thru 40 HP	25	25	45	45	45
Base Mounted 50 HP & Over	25	25	45	65	65
Packaged Air Handling Equipment					
Suspended Up to 5 HP	25	25	25	25	25
Suspended 7.5 HP & Over:					
Up to 500 RPM	25	45	45	45	45
501 RPM & Over	25	25	25	45	45
Floor Mounted Up to 5 HP	10	25	25	25	25
Floor Mounted 7.5 HP & Over:					
Up to 500 RPM	10	45	45	45	45
501 RPM & Over	10	25	25	45	45
Fans & Blowers					
Utility Sets Up to 500 RPM	10	45	45	45	65
Utility Sets 501 RPM & Over	10	25	25	45	45
Fan Heads & Tubular Fans Up to 50 HP:					
Up to 300 RPM	10	65	65	65	90
301-500 RPM	10	45	45	65	65
500 RPM & Over	10	25	25	45	65
Fan Heads & Tubular Fans 60 HP & Over:					
301-500 RPM	25	45	50	65	90
501 RPM & Over	25	45	25	45	65
Axial Fans Up to 50 HP	25	25	25	45	65
Axial Fans 60 HP & Over	25	25	45	65	90
Centrifugal Blowers Up to 50 HP:					
Up to 200 RPM	10	65	65	90	90
201 to 300 RPM	10	45	65	65	90
301 to 500 RPM	10	45	45	65	90
501 RPM & Over	10	25	25	45	90
Centrifugal Blowers 60 HP & Over:					
Up to 300 RPM	45	65	90	90	90
301 to 500 RPM	45	45	65	90	90
501 RPM & Over	25	45	45	65	65
Cooling Towers					
Up to 500 RPM	10	10	45	65	90
501 RPM & Over	10	10	25	45	65
Internal Combustion Engines					
Up to 25 HP	10	10	45	65	65
30 to 100 HP	10	45	65	90	90
125 HP & Over	10	65	90	120	120

Fig 4

Chapter 5
Installation

Once a resilient mount has been selected, either through a procedure evolved from a desired transmissibility or from a more broadly based quick selection chart (Chapter 4, Fig 4), it is necessary to study how it is best employed and take note of any subsidiary advantages or snags which may result from a selected applications procedure.

Some of the necessary points for attention are listed here:

1.   Mount type - spring, rubber, glass fibre

2.   Load rating - catalogue

3.   Location of mount on equipment

4.   Strain on equipment frame from limited (eg 4) locations

5.   Steel base frames

6.   Concrete inertia base frames

7.   Flexible couplings

8.   Pipe couplings and resilient suspension

9.   Fan reaction pressures

Fig 1 shows additional features added to those presented in the previous chapter as it now includes an indication as to whether spring (2), rubber or glass fibre (1) mounts are required. Also is included an indication as to whether a steel base frame (5) & (6) or a concrete inertia base frame (7) is required. The symbol (3) indicates a restrained coiled spring. This considerably assists with points 1, 5 and 6, as will be seen later.

Mount Location

There are two approaches to mount location; the first consisting of choosing a convenient fixing position, such as the four corners of an inertia block, and then selecting mountings of different ratings to carry the loads that occur at these selected points, taking into account the required deflection in conjunction with a catalogue selection; alternatively, positions may be selected such that the mounts are all equally loaded and catalogue selection is correspondingly simplified.

There are distinct advantages in adopting the latter approach of mount equality as it obviously simplifies mount fixing on site as no thought is required as to which mount goes at which location. However, there are often problems in being able to select convenient mounting locations and the mathemetics of selection is sometimes overcomplicated. We will see later how the employment of inertia bases can ease the situation leading to similar mount loadings while making full use of the height adjustment screw in conjunction with the tolerance of such mounts which usually exceeds at least 30%. Low deflection mounts, more especially of the fibreglass variety., are more forgiving in this aspect of critical selection than are the high deflection spring mounts.

Four Mounts

The simplest situation is met when the centre of gravity lies obviously at the geometric centre of the four mounting points when each resilient isolator is selected to carry $\frac{1}{4}$ of the total mass.

However, if as is most likely, the arrangement is not symmetric, the loading is uneven as in Fig 2. In this case the loading is derived by taking moments resulting in the calculations indicated on Fig 2. We will see later a simple quick selection procedure enabling us in practice to avoid the tedium of such a calculation.

Equal Loading

Even loading can be restored to this situation in Fig 2 in one direction by providing a pair of intermediate mounts as shown in Fig 3. Here for simplicity the centre of gravity has been located centrally in one direction while remaining offset in the other. The location of the two additional mounts has again been derived by taking moments but this time each of the six mounts will support equally 1/6th of the total weight. The calculation is indicated on Fig 3.

However, it is sometimes necessary to employ more distributed mounts as a substantial frame is not inherent in the equipment construction, eg air handling unit. This previously would have been conventionally supported on a continuous floor slab, requiring little strength from the frame. Sufficient point locations are thus necessary to simulate this continuous support. Fig 4 illustrates this situation for ten mounts.

For less regular frame designs such as centrifugal fans, scrolls and motor combinations the solution for a five and six mount selection is indicated in Figs 5 and 6.

Having adopted this procedure it must be remembered that weight transfer is occurring via the frame structure and it must be ensured that this frame is able to transfer this load to the corresponding point locations. Inertia bases and steel frame bases are often separately recommended in order to ensure this facet of selection.

Fan Reaction Pressures

Having placed our piece of equipment on resilient mounts it is obviously more prone to movement, both dynamic as a result of vibration, and static displacements as a result of any subsidiary reacting forces. With high pressure, high velocity fans, mainly of centrifugal design, there is considerable back reaction on the fan as it reacts against the static and velocity pressure heads. Owing to the necessary flexible coupling on the outlet of such a fan this reaction, often occurring at some distance from the base, will tend to tilt the base as a result of its turning moment (Fig 7). The magnitude of this problem is indicated in Fig 8 which incorporates a high pressure fan working at 200mm (8in) swg with a discharge area of 625mm x 500mm (25in x 20in). It can be seen that the turning moment due to reaction pressure is about half of that due to the static weight of the machine. To overcome this problem the fan may be pre-tilted as shown in Fig 9 or an inertia block employed as in Fig 10.

The use of the inertia block will require stiffer springs to yield the same deflection from the increased weight and the stiffer springs will be more able to resist the turning motion. Fig 10 illustrates the addition of an 800kg (1800lb) inertia block to the fan of Fig 8 which adds static mass to the system and increases the separation of the mounts. The moment of the reaction pressure is now reduced to only 1/12th of that due to the static weight. Should it prove impractical to employ an inertia base and high deflection springs have been requested, the pre-tilted situation of Fig 9 must be employed in conjunction with snubbers to prevent the off load tilt from straining any flexible connectors.

Inertia Bases

Inertia bases are employed for a variety of reasons and some of the advantages gained are listed here:

1.   To give a more even weight distribution for the resilient mounts

2.   To minimise the effects of errors in the position of the equipment's centre of gravity.

3. To give more stability to a system by lowering its centre of gravity

4. To provide rigidity to the equipment assembled parts

5. To reduce tilt on high pressure fans

6. To limit the amplitude of motion when running up or down through resonance

7. To act as a local barrier to direct noise impinging on the floor slab

Some of these features are also supplied by steel frame bases which offer some additional mass but mainly rigidity.

Inertia bases are particularly useful for machines with large out of balance forces such as internal combustion engines, air compressors, reciprocating fridge compressors, pumps and high pressure fans although the justification for their employment is often different for each machine. As has been mentioned earlier doubling the effective mass of the machine with an inertia block will halve the amplitude of vibration of the machine. This is of special significance at resonance, Fig 11.

With Pumps: It can be seen from Table 1 that inertia blocks are called for with many varieties of pumps even though centrifugal in action. Most often they are recommended to supply rigidity between the components of base mounted units where the motor and pump are separately presented and linked by a drive shaft or belt. This is illustrated in Fig 12. Such a rigid base is usually chosen with a low mass ratio between the inertia block and the machine in the region of 1:1. The improved rigidity considerably extends the life of the compounded assembly and is most useful in reducing pump motion. This reduced motion is of assistance with subsequent pipe connections and isolation. Generally the low powered close coupled pumps are simply mounted without an inertia block with rubber mounts supplying acoustic isolation and avoiding ground resonances above 20Hz.

Inertia blocks may be arranged to be only under the pump or have extended length to also support under the inlet elbow of "End Suction" pumps as contrasted in Fig 13.

With Internal Combustion Engines: Here the machine is basically out of balance and the vibration amplitudes resulting simply from the employment of the requested resilient mount deflection will lead to considerable motion of the engine. It is usually mounted in conjunction with an inertia block of mass ratio up to 6:1 in order to reduce this vibration amplitude and hence lead to considerably enhanced engine life. It will be noticed that the degree of isolation required in a basement situation is still quite low employing rubber or fibreglass mounts to supply acoustic isolation.

With Fans: In low pressure applications (0-75mm (0-3in) swg) it will be
seen that with floor mounted equipment employing coil springs, steel frame
bases are called for in Table 1.   This is simply to improve the rigidity
of the construction when soft coil springs are employed and thus enables
more simple application of four mounts to the convenient corners of this
base frame, Fig 14.

However, with high pressure units (75-250mm (3-10in) swg) inertia bases
may prove necessary for reasons of "tilt" control when coil springs are
employed as previously described.   On such bases there should not be any
problems of sideways thrust up to 200mm (8in) swg with open coil springs
because of the increased stiffness necessary vertically to support the base
which will be accompanied by a corresponding stiffening horizontally.   Any
flexible connection will then only be subjected to running displacements of
about 3mm (1/8in) and transient run-up resonance movements under 12mm
($\frac{1}{2}$in).

The mass ratio chosen for fan applications should be about 3:1 for high
pressure units but later it will be shown that this mass is usually determined,
and achieved, from inertia base design criteria.

With low pressure centrifugal fans (10-75mm (0-3in) swg) extra mass is not
usually necessary, but sufficient rigidity must be in the beam base to enable
a simple 4 to 6 mount point location system to be employed.   Modest bending
or twisting of a base frame - more usually supported continuously on a floor
slab - can lead to mysterious accelerated wear rates.   A subsidiary steel
beam base should be employed.

With Reciprocating Compressors: As with internal combustion engines,
large out of balance forces are produced by most reciprocating air compressors.
These forces need adequate isolation, even in basement areas, and coil springs
are specified to have at least 25mm (1in) deflection.   This simple mounting
configuration will then usually result in large vibrational movements of the
machine.   Sometimes these are reduced sufficiently by the incorporation of
the compressor on the storage cylinder and this supplies the appropriate mass
and inertia.   If the compressed air is fed away by flexible rubber hosing
this vibration amplitude may be acceptable and of no inconvenience.

However, for a well mounted unit a heavy inertia base should be employed
with a mass ratio of about 6:1, to reduce this vibration down to at most
3mm (1/8in).

Mount Location

Inertia Blocks - Having employed an inertia block or steel base frame, sufficient rigidity will now be present to allow the mounts - usually 4 to 6 - to be simply placed at the corners and symmetrically along the sides Fig 15.

The spring rates of each resilient mount will usually be different and a clear marking scheme is essential. Evaluation of these rates and loadings will be dealt with under "Quick Selection Method".

General Comment: It will be noticed in Table 1 that many machines call for inertia bases when higher deflection coil springs are required. This is usually to supply inertial and tilt control to the system now that larger mount deflections are required.

Axial Fans

In low pressure applications axial flow fans do not usually require bases, as their inherent cylindrical design gives them adequate rigidity. Mount selection is usually the same as for centrifugal fans as either minimal acoustic isolation is required or the avoidance of loaded structural floor resonances leads to the selection of coil springs. However it is fair to add that due to their good mechanical and aerodynamic balance, the higher speed units 47Hz (2800rpm) and 23Hz (1400rpm) can usually be employed with 9mm (0.35in) deflection rubber or glass fibre mounts on 6m (20ft) floor spans.

When cylindrical attenuators are employed with an axial flow fan - as frequently is the case - they should be bolted directly to the fan and the total unit supported on four pairs of resilient mounts as shown in Fig 16. This ensures optimum aerodynamic and acoustic operation of the fan and adds useful inertia to the resilient assembly. Flexible duct connectors are then still required at the ends of attenuators as attenuators are not "vibration isolators" but only ductborne noise attenuators.

Centrifugal Chillers

Only the open centrifugal type of chiller requires an inertia base and then for reasons of machine base rigidity, exactly as with pumps, rather than for inertial mass control.

However when higher deflection coil springs are required with the sealed unit, steel rails are used to separate the mounts and yield adequate stability. The rails are placed across the narrow dimension and chosen of a length to form an imaginary equilateral triangle between the two mount locations and the chillers centre of gravity, Fig 17.

This triangle is a good design code to adopt on any beam base arrangement.

Cooling Towers

Beams or open beam bases will be required on this equipment if integral base frames are not adequate. Then simple mount location is usually restricted to the four corners, employing rubber, fibreglass or height restrained coil springs.

QUICK SELECTION METHOD

Four Mount System

In order to mathematically calculate the weight distribution on each mount of a four mount isolated system, eg (1) a fan and motor mounted on a grillage (2) a cooling tower mounted on a steel frame base (3) a diesel generator mounted on a concrete inertia base, it is necessary to run moments in two directions. This is time consuming and mistakes may easily be made in the arithmetic.

Six Mount System

Steel frame bases and concrete inertia bases usually have six mounts equally spaced (three on each side). It is these six mount systems referred to here. In order to mathematically calculate the weight distribution on each of the equally spaced mounts it is necessary to run moments in two directions. The mathematics required to produce a solution is considerable and even then the solution obtained is only a general one - there is not a unique solution.

The reason for this is because there is an infinite number of ways in which the loading can be spread on an equally spaced six mount system. For example, at one extreme the two centre mounts could take the total weight of the system leaving no weight to be carried by the four outer mounts. At the other extreme the four outer mounts could carry all the system weight leaving no weight for the two centre mounts to carry. In either of these situations a large amount of strain is set up in the base (whether steel frame or inertia) due to bending moments. Between these two extremes there must be a loading situation such as to give the least amount of strain in the base.

The Quick Selection Method

The selection method put forward here not only deals with the four mount system to a practical degree of accuracy, but also gives the "least base strains" solution which is ideally required for the six mount system.

Referring to Fig 18, we have a matrix of large dots which represent positions of centres of gravity. Around each large dot are grouped four numbers which represent the percentage distribution of weight at each of the four corners of an isolated system due to the weight acting through that particular position of centre of gravity.

Along the x and y axis are numbers 1 to 10 which represent a scale of base, length and width.

In a practical situation it is necessary to split up the base in question into ten equal parts in the length and width, and knowing the positions of the centres of gravity of each component of the mounted system, these are compared with the equivalent positions on the matrix diagram and their percentage weight distribution on each of the four corner mounts is established. This is done for each component, (motor, fan, concrete, beams etc.), as if separate and their contributions at each corner summed.

Example of a four mount system calculation:

Fig 19 shows a concrete inertia base supporting a fan and motor. The positions of the centre of gravity of the fan and motor are indicated and the centre of gravity of each beam is shown halfway along its length. The centre of gravity of the concrete slab is shown at the centre of the slab.

Lines are drawn such that the base drawing is split into ten equal parts in the length and width. The centres of gravity may then be compared with the standard matrix and the % weight distributions read off.

Let us assume weight as follows:

Fan weight	500kg
Motor weight	200kg
Weight of beam AB	40kg
Weight of beam CD	40kg
Weight of beam AC	30kg
Weight of beam BD	30kg
Weight of concrete	1500kg
Total base weight	2340kg

Now compare centres of gravity with the standard matrix and obtain the following table:

Component	Weight Distribution on Each Mount			
	A	B	C	D
Fan	35% of 500 = 175kg	15% of 500 = 75kg	35% of 500 = 175kg	15% of 500 = 75kg
Motor	4% of 200 = 8kg	16% of 200 = 32kg	16% of 200 = 32kg	64% of 200 = 128kg
Beam AB	50% of 40 = 20kg	50% of 40 = 20kg	0% of 40 = 0kg	0% of 40 = 0kg
Beam CD	0% of 40 = 0kg	0% of 40 = 0kg	50% of 40 = 20kg	50% of 40 = 20kg
Beam AC	50% of 30 = 15kg	0% of 30 = 0kg	50% of 30 = 15kg	0% of 30 = 0kg
Beam BD	0% of 30 = 0kg	50% of 30 = 15kg	0% of 30 = 0kg	50% of 30 = 15kg
Concrete	25% of 1500 = 375kg	25% of 1500 = 375kg	25% of 1500 = 375kg	25% of 1500 = 375kg
Total weight	593kg	517kg	617kg	613kg

Giving a total weight on the four mounts of 2340kg which checks with the total base weight.

Example of a six mount system calculation:

As previously stated, this selection method ensures that the base will be subjected to the least possible strain.

The method involves treating the six mount system as two four mount systems, by splitting the base down the line of the two centre mounts.

Fig 20 shows a typical steel frame base with six mounts. This is split into two as shown on the drawing and then the procedure is the same as for the four mount systems. When the weight distribution on all eight mounting points has been established the centre mounting points are combined such that Mount E = mounting point G + H and Mount F = mounting point J + K.

It may be in some cases with a six mount system that the machine (say a chiller) could lie across the length of the base such that when the imaginary "split" across the two centre mounts is made the machine is also "split". In this case the "split" weight of each half of the machine must be known and its approximate centre of gravity on each half of the base.

If this information is not known it is possible to establish the weight of each half of the machine acting on each part of the base by a simple moments calculation.

For example:

Fig 21 shows a six mount system with a piece of equipment mounted along its length. We will split the base at the centre mounts and for convenience we will calculate the weights acting halfway along the length of each base.

Taking moments about A we have:

$460kg \times 600mm = W_2 \times 2400mm$ ($1000lb \times 2ft = W_2 \times 8ft$)

Taking moments about B we have:

$460kg \times 1800mm = W_1 \times 2400mm$ ($1000lb \times 6ft = W_1 \times 8ft$)

Therefore: $W_2 = 115kg$ ($250lb$)
$W_1 = 345kg$ ($750lb$)

We can now treat this system as two four mount systems as explained in the previous example of a six mount system.

Installation

Equipment Type	On Grade			On 6m Floor Span			On 9m Floor Span			On 12m Floor Span			On 15m Floor Span		
	Base Type	Isolation Type	Min. Defl. mm	Base Type	Isolation Type	Min. Defl. mm	Base Type	Isolation Type	Min. Defl. mm	Base Type	Isolation Type	Min. Defl. mm	Base Type	Isolation Type	Min. Defl. mm
Refrigeration Machines															
Absorption	4	1	6	4	3	25	4	3	25	4	3	45	4	3	45
Packaged Hermetic	4	1	10	4	3	25	4	3	45	4	3	45	5	3	65
Open Centrifugal	6	1	10	6	3	25	6	3	45	6	3	45	6	3	90
Reciprocating:															
500-750 RPM	4	3	25	4	3	45	5	3	45	5	3	65	5	3	90
751 RPM & Over	4	3	25	4	3	25	4	3	45	5	3	65	5	3	65
Compressors															
Air or Refrigeration															
500-750 RPM	4	2	25	4	2	45	7	2	65	7	2	65	7	2	90
751 RPM & Over	4	2	25	4	2	25	7	2	45	7	2	65	7	2	65
Boilers or Steam Generators															
All	4	1	6	4	3	25	4	3	25	4	3	45	4	3	65
Pumps															
Close Coupled thru 5 HP	4	1	10	7	2	25	7	2	25	7	2	25	7	2	25
Close Coupled 7.5 HP & Over	7	2	25	7	2	25	7	2	25	7	2	45	7	2	45
Base Mounted thru 40 HP	7	2	25	7	2	25	7	2	45	7	2	45	7	2	45
Base Mounted 50 HP & Over	7	2	25	7	2	25	7	2	45	7	2	65	7	2	65
Packaged Air Handling Equipment															
Suspended Up to 5 HP	4	2	25	4	2	25	4	2	25	4	2	25	4	2	25
Suspended 7.5 HP & Over:															
Up to 500 RPM	4	2	25	4	2	45	4	2	45	4	2	45	4	2	45
501 RPM & Over	4	2	25	4	2	25	4	2	25	4	2	45	4	2	45
Floor Mounted Up to 5 HP	4	1	10	4	2	25	4	2	25	4	2	25	4	2	25
Floor Mounted 7.5 HP & Over:															
Up to 500 RPM	4	1	10	5	2	45	5	2	45	5	2	45	5	2	45
501 RPM & Over	4	1	10	4	2	25	4	2	25	5	2	45	5	2	45
Fans & Blowers															
Utility Sets Up to 500 RPM	4	1	10	4	2	45	5	2	45	5	2	45	5	2	65
Utility Sets 501 RPM & Over	4	1	10	4	2	25	4	2	25	5	2	45	5	2	45
Fan Heads & Tubular Fans Up to 50 HP:															
Up to 300 RPM	4	1	10	5	2	65	5	2	65	5	2	65	5	2	90
301-500 RPM	4	1	10	5	2	45	5	2	45	5	2	65	5	2	65
500 RPM & Over	4	1	10	4	2	25	4	2	25	5	2	45	5	2	65
Fan Heads & Tubular Fans 60 HP & Over:															
301-500 RPM	5	2	25	5	2	45	5	2	50	5	2	65	5	2	90
501 RPM & Over	5	2	25	5	2	45	5	2	25	5	2	45	5	2	65
Axial Fans Up to 50 HP	4	2	25	4	2	25	4	2	25	5	2	45	5	2	65
Axial Fans 60 HP & Over	5	2	25	4	2	25	4	2	45	7	2	65	7	2	90
Centrifugal Blowers Up to 50 HP:															
Up to 200 RPM	6	1	10	6	2	65	6	2	65	6	2	90	6	2	90
201 to 300 RPM	6	1	10	6	2	45	6	2	65	6	2	65	6	2	90
301 to 500 RPM	6	1	10	6	2	45	6	2	45	6	2	65	6	2	90
501 RPM & Over	6	1	10	6	2	25	6	2	25	6	2	45	6	2	90
Centrifugal Blowers 60 HP & Over:															
Up to 300 RPM	6	2	45	7	2	65	7	2	90	7	2	90	7	2	90
301 to 500 RPM	6	2	45	7	2	45	7	2	65	7	2	90	7	2	90
501 RPM & Over	6	2	25	7	2	45	7	2	45	7	2	65	7	2	65
Cooling Towers															
Up to 500 RPM	4	1	10	4	1	10	5	3	45	5	3	65	5	3	90
501 RPM & Over	4	1	10	4	1	10	5	3	25	5	3	45	5	3	65
Internal Combustion Engines															
Up to 25 HP	7	1	10	7	1	10	7	2	45	7	2	65	7	2	65
30 to 100 HP	7	1	10	7	2	45	7	2	65	7	2	90	7	2	90
125 HP & Over	7	1	10	7	2	65	7	2	90	7	2	120	7	2	120

Isolator Types:
1. Rubber or glass fibre floor isolators, or hangers.
2. Free-standing spring floor isolators, or hangers.
3. Restrained spring isolation.

Base Types:
4. No base. Isolator attached directly to machine.
5. Structural rail base, used to reduce mounting height.
6. Completely fabricated structural rail base, i.e., for fans and motors.
7. Concrete inertia base.

Fig 1

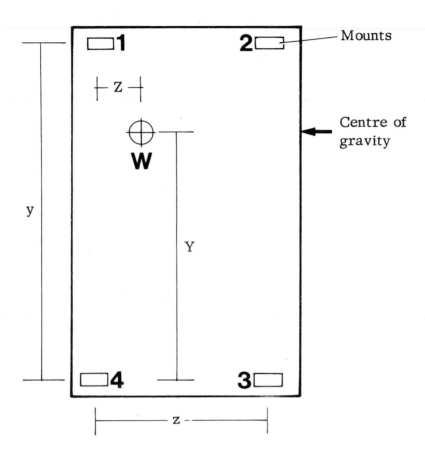

Let W represent the weight of the object concentrated at the centre of gravity and 1, 2, 3 and 4 the mounts on which the load rests. The position of the centre of gravity with respect to the mounts is given by the dimensions y, Y, z and Z.

The result, by taking moments, is given below:

Load on mount 1 $= W \times \dfrac{(z - Z)}{z} \times \dfrac{Y}{y}$

Load on mount 2 $= W \times \dfrac{Z}{z} \times \dfrac{Y}{y}$

Load on mount 3 $= W \times \dfrac{Z}{z} \times \dfrac{(y - Y)}{y}$

Load on mount 4 $= W \times \dfrac{(z - Z)}{z} \times \dfrac{(y - Y)}{y}$

Fig 2   Uneven four point mounting

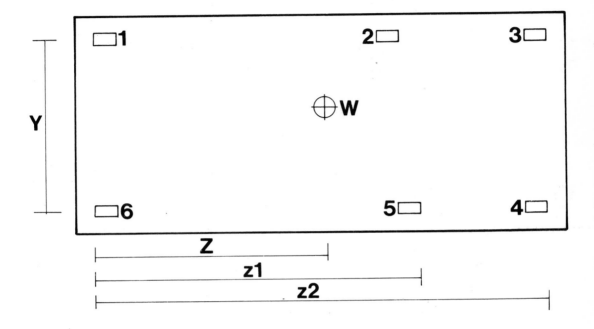

If the centre of gravity is located centrally in one direction but offset in the other as shown then two more mounts 2 and 5 may be added at a distance z_1 as shown, such that all six mounts support the same weight - that is $\dfrac{W}{6}$

To calculate z_1 take moments about the line joining mounts 1 and 6

Thus $\dfrac{W}{3} \times z_1 = W \times Z - \dfrac{W}{3} z_2$

So $z_1 = 3Z - z_2$

Fig 3   Six point mounting, centre of gravity offset to one side

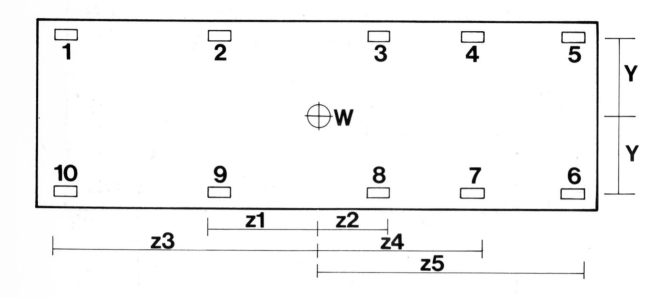

The weight resting on each mount will be $\frac{W}{10}$ if the following condition is met:

$z_3 + z_1 = z_2 \times z_4 + z_5$ when Y = Y

That is, clockwise moments equal anticlockwise moments.

Fig 4   Many mounts arranged for equal weight distribution, one direction only considered

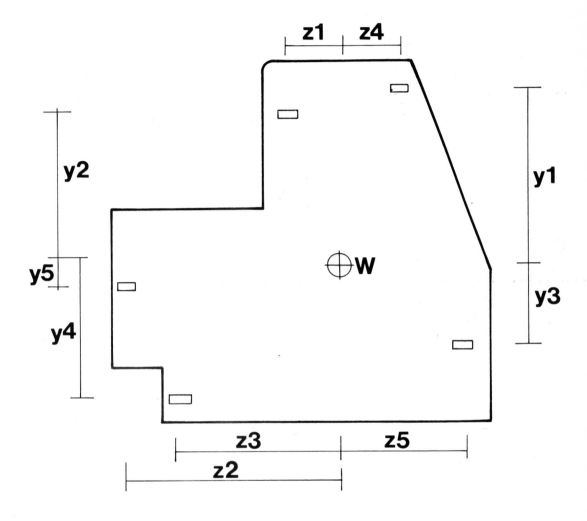

There are five mounts arranged about the centre of gravity. If we take moments in two directions at right angles about this centre of gravity we arrive at the two conditions given below if we assume that the weight resting on each mount is $\frac{W}{5}$.

Then $z_1 + z_2 + z_3 = z_4 + z_5$

and $y_1 + y_2 = y_3 + y_4 + y_5$

Fig 5   General case, many mounts arranged for equal weight distribution, two directions considered

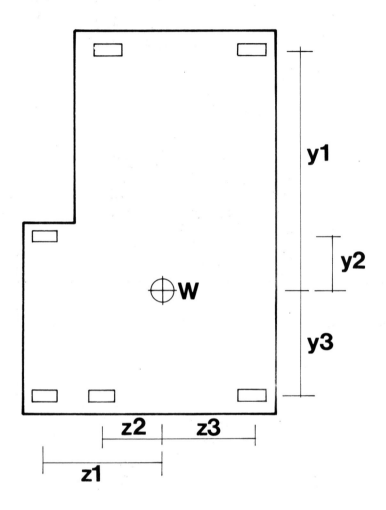

The weight per mount will be $\dfrac{W}{6}$ if the following conditions are met:

$$z_1 + z_2 = z_3$$

and

$$2y_1 + y_2 = 3y_3$$

Fig 6   Particular case of above - but with six mounts

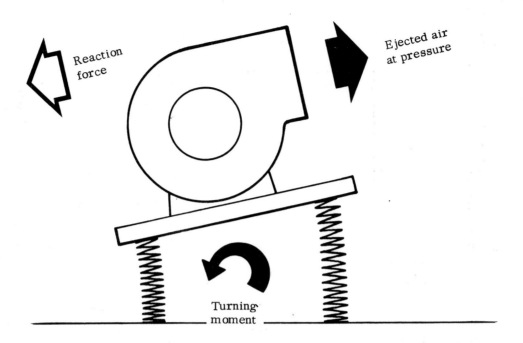

Fig 7   Fan reaction pressures

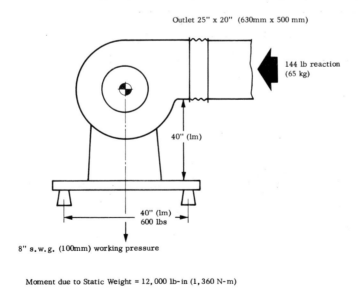

Outlet 25" x 20" (630mm x 500 mm)

144 lb reaction
(65 kg)

40" (1m)

40" (1m)
600 lbs

8" s.w.g. (100mm) working pressure

Moment due to Static Weight = 12,000 lb-in (1,360 N-m)

Moment due to Reaction Pressure = 5,760 lb-in (650 N-m)

Fig 8

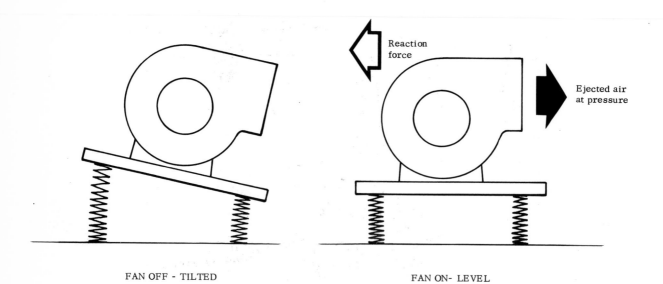

FAN OFF - TILTED

FAN ON- LEVEL

Fig 9   Pre-tilted fan

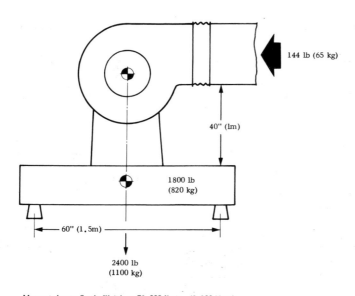

Moment due to Static Weight = 72,000 lb-in. (8,200 N-m).

Moment due to Reaction Pressure = 5,760 lb-in. (650 N-m).

Fig 10

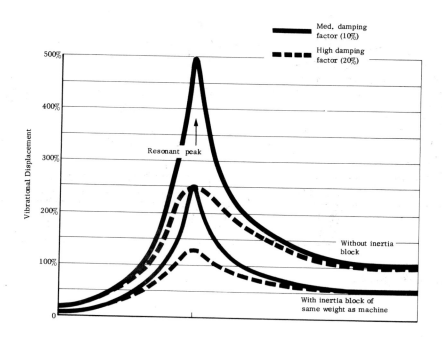

Fig 11   Inertia block control

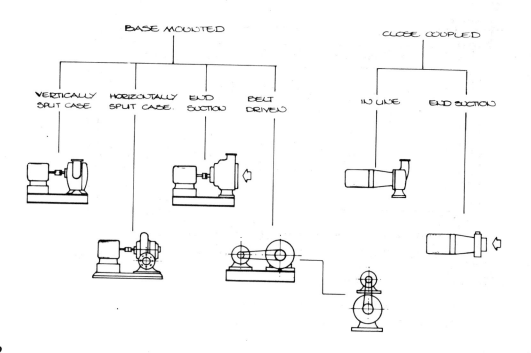

Fig 12

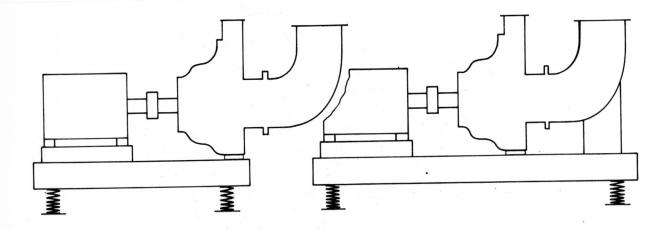

Fig 13   Elbow support

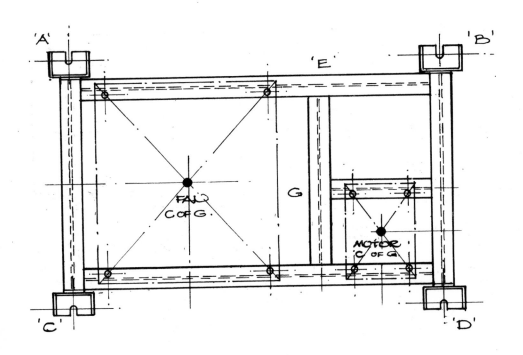

Fig 14

Fig 15   Inertia base before concreting

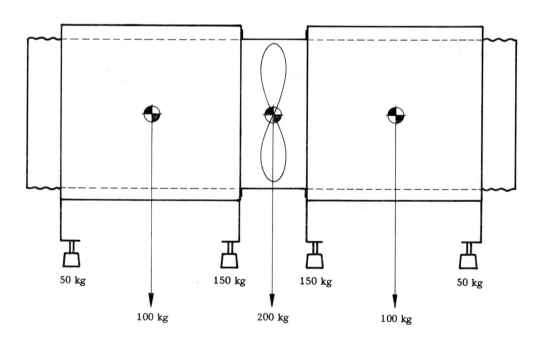

50 kg   150 kg   150 kg   50 kg

100 kg   200 kg   100 kg

Fig 16   Axial fan + attenuators

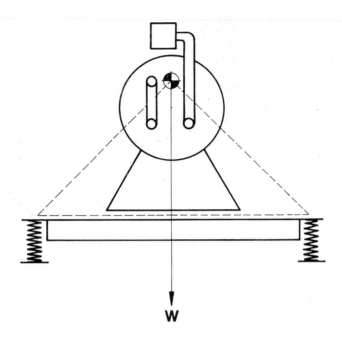

Fig 17   Beams to improve stability

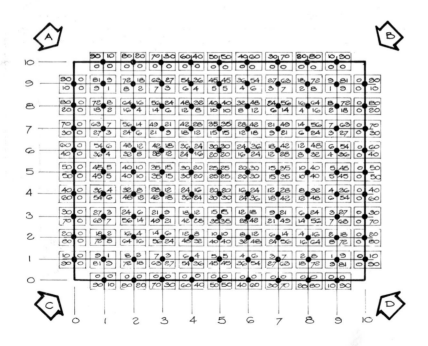

Fig 18

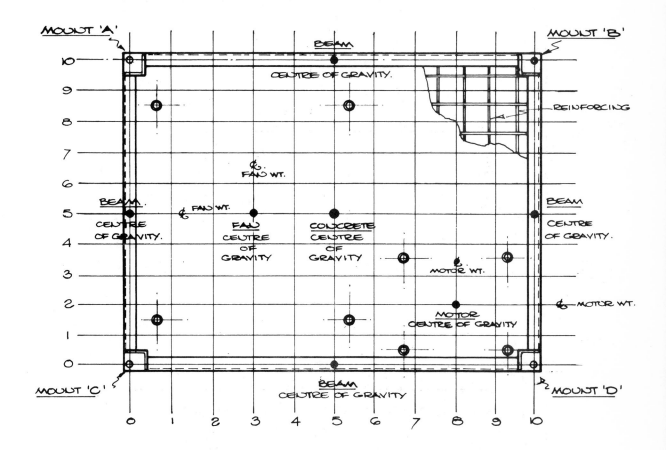

Fig 19

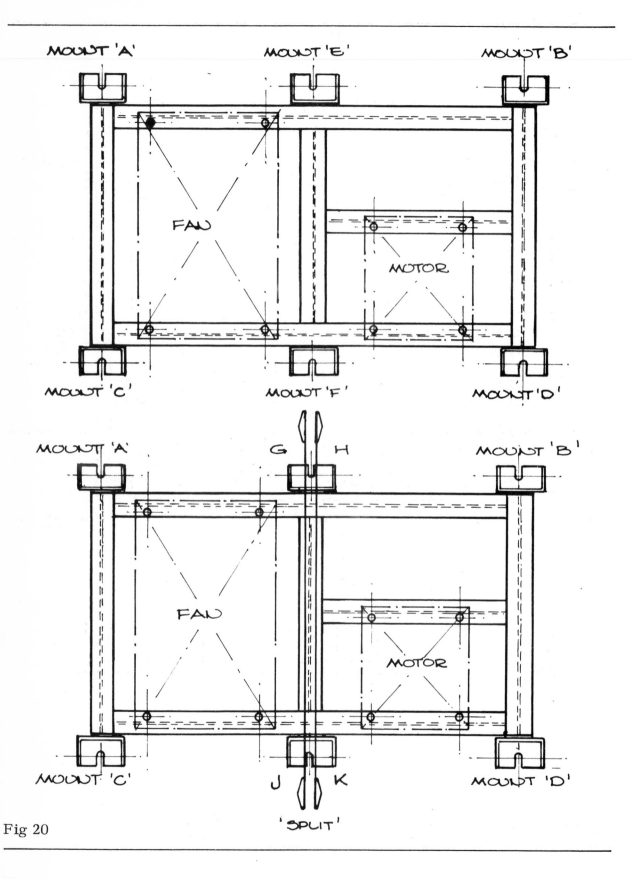

Fig 20

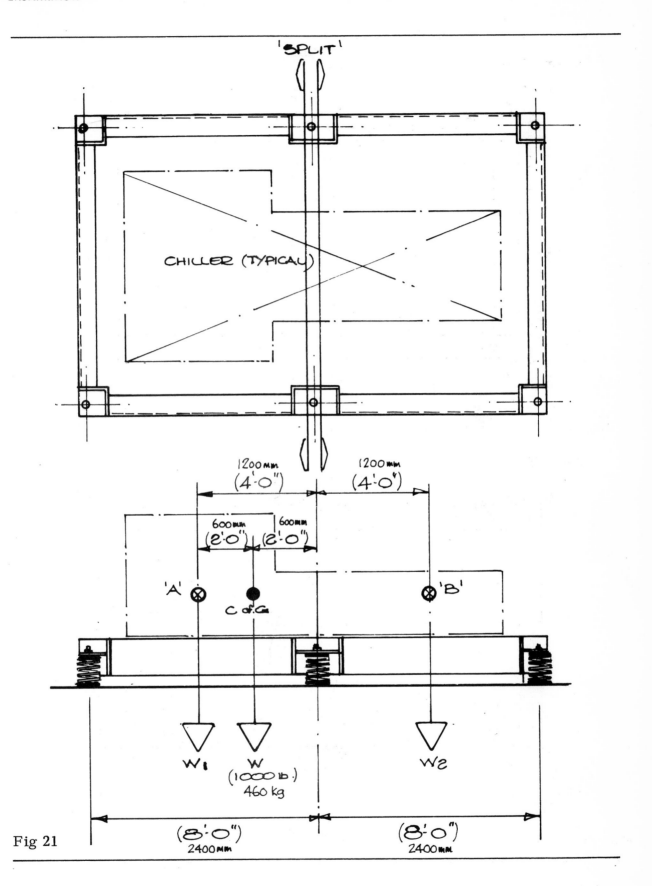

Fig 21

Chapter 6
Plantroom Design and Services Installation

1. Introduction

Any single unit of mechanical services equipment installed in a building can be considered to interface with the building structure in one or more of the following ways:

1.1 Physical support of the dead load by a slab or framework
1.2 Connection to remote areas by continuous ducts and pipework
1.3 Connection to sources of power such as mechanical drive assemblies, electrical connections, fuel lines, etc.
1.4 Connections to ancillary equipment such as controls and linked systems
1.5 Connections to waste product disposal systems such as flues or drains

Consequently if any item of mechanical services plant is a potential source of vibration excitation, it is the <u>linked</u> system that must be viewed in its entirety and vibration must be decoupled at all points where potential transmission of unwanted energy transfer could occur.

The basic rules for the control of vibration follow on from the methods discussed in previous chapters. In general, the decoupling will be achieved by introducing a vibration isolating discontinuity between the energy source and all points of interface with the building and the connected services. Alternatively it is possible to allow natural dissipation of vibrational energy along service links but to introduce decoupling at all mounting or attachment points (see Fig 1). Since most services which are linked to a source of vibration are subject to the following constraints:

1.6 their distributed weight must be supported by load bearing structures
1.7 they must be physically located in precise positions with relation to other equipment and fittings
1.8 they will be linked to other services
1.9 they will occasionally penetrate walls and slabs

a variety of techniques may have to be adopted in order to introduce a vibration decoupling discontinuity at such points. Table 1 illustrates standard methods of successfully decoupling transmitted vibration at points in linked systems.

Many items of mechanical services plant are fairly crude in construction with rotor imbalances present to an appreciable extent in some cases. It is desirable that all rotors and impellers should be balanced to the advised limits (this preferably being done on site) as the movement of any machine on its anti-vibration mountings when operating will be dependent on its state of balance (see Figs 2 and 3 and Table 2).

VIBRATION TRANSFER ROUTE	SUPPORTS IN COMPRESSION	SUPPORTS IN TENSION	LINK	BUILDING PENETRATION
Machine mountings	springs, compliant pads, rubber in shear, rubber in compression, 'floated floors'	spring hangers 'inverted' anti-vibration mounts	-	-
Ductwork	pedestal pads, anti-vibration mountings	"	flexible collar	resilient packing
Flue	high-temperature pads, steel springs	steel spring hangers	asbestos collar	vibration isolation not advised at this point
Steam pipe	"	"	metal bellows	fibrous asbestos sleeve
Water pipe	compliant pads, anti-vibration mounts, springs	coil springs or anti-vibration hangers	metal bellows rubber bellows braided hose	resilient packing anti-vibration collar or sleeve
Fuel or Vacuum Pipes	"	"	looped metal tube braided hose	"
Refrigerant pipe	"	"	braided flexible metal hose	"
Electrical cable	cable tray on compliant pads	cable tray on anti-vibration hangers	flexible conduit hanked cable	none required
Rotating shaft	-	-	flexible coupling	-
Drain	-	-	discontinuity, hose	resilient sleeve

Table 1

Quality Grade G	Velocity of displacement of c. of g. mm/s	Rotor types – General examples
G 4000	4000	Crankshaft-drives of rigidly mounted slow marine diesel engines with uneven number of cylinders.
G 1600	1600	Crankshaft-drives of rigidly mounted large two-cycle engines.
G 630	630	Crankshaft-drives of rigidly mounted large four-cycle engines. Crankshaft-drives of elastically mounted marine diesel engines.
G 250	250	Crankshaft-drives of rigidly mounted fast four-cylinder-diesel engines.
G 100	100	Crankshaft-drives of fast diesel engines with six and more cylinders. Complete engines for cars, trucks and locomotives.
G 40	40	Car wheels, wheelrims, wheel sets, drive shafts. Crankshaft drives of elastically mounted fast four-cycle engines with six and more cylinders. Crankshaft-drives for engines of cars, trucks and locomotives.
G 16	16	Parts of crushing machinery. Individual components of engines for cars trucks and locomotives.
G 6.3	6.3	Centrifuge drums. Fans. Aircraft gas turbine rotors. Fly wheels. Pump impellers. Machine-tool and general machinery parts. Normal electrical armatures.
G 2.5	2.5	Gas and steam turbines. Turbo-compressors. Machine-tool drives. Small electrical armatures.
G 1	1	Tape recorder and gramophone drives. Grinding-machine drives.
G 0.4	0.4	Precision grinders. Gyroscopes.

Table 2   Balancing grades for various groups of representative rigid rotors

Where possible flexible discontinuities in services systems should:

1.10.   isolate all significant excitation frequencies
1.11.   be located at a point of minimum disturbance
1.12.   have less stiffness than the main mounts
1.13.   prevent break-out of any contained sound fields
1.14.   withstand transient start-up motion
1.15.   control any rocking of the machine
1.16.   withstand any strain imposed on the connectors
1.17.   not limit the freedom of any machine to move within established limits

2. Flexible Connectors in Ducts

In general, vibration excitation of ductwork will be caused by imbalance of the prime mover (fans and blowers) or motion of the duct walls induced internally by turbulent airflow. It is standard practice to decouple ductwork vibration by means of discontinuities such as flexible collars fabricated from canvas, plastic or sound barrier mat. Any material used for flexible joints in ductwork must withstand the specified conditions of temperature and air pressure and must comply with the necessary standards of air-tightness specified for the rest of the duct system. The material should be fire resistant and where necessary corrosion and abrasion resistant. Woven-fibre and lightweight plastic materials provide limited sound insulation and noise breakout from such flexible joints between fans and their associated ductwork could give rise to excessive noise levels in surrounding areas. It is therefore recommended that flexible sound-insulating materials be used generally for fan connectors and throughout high pressure ducted systems. Any material selected for use as a sound-insulating flexible collar should have a weight of at least $5kg/m^2$ and should remain flexible over the anticipated temperature range for the contained air flow. It should be fixed to associated ductwork by hose clips or metal or fibre banding supplemented, where necessary, by adhesives.

Care should be taken during the installation of flexible collars to ensure that they do not sag under negative pressure and restrict the contained air flow, particularly on the upstream side of axial flow fans, where excess noise generation could occur if the impeller blade tips run in a turbulent boundary layer. When flexible duct connectors are used on the high pressure side of fans, care should be taken to ensure that extension of the flexible joint does not cause a movement of the fan or associated equipment. This point should be carefully checked where large flexible connectors are used on the downstream side of high pressure centrifugal fans. Any bellows-type extension of this joint under pressure will tend to cause an over-turning moment at the fan on its anti-vibration mountings - particularly where AVM's having static deflections in excess of 25 mm have been specified for the air-moving equipment.

A clearance of at least 10 mm should always be allowed between duct spigots connected by a flexible collar in order to permit transient movement of the machine during start-up.

3. Vibration Isolating Pipes and Conduits

This classification of service connectors includes all circular-section narrow bore tubes transporting liquids and gases (usually under positive pressure), partial vacuum systems and conduit or containment for electrical systems. Generally speaking all such pipework has relatively thick rigid walls which can transmit machine vibration over long distances. Pipes seldom radiate high noise levels from their surfaces, but when coupled to a large radiating surface such as a slab or partition the resulting excitation of the larger "sounding board" can give rise to high levels of perceived sound.

Pipe borne vibration will usually be found to originate at either a rotational source or a reciprocating source. Rotational sources such as centrifugal pumps, vane or screw impellers etc. give rise to a complex sound spectrum made up from some or all of the following components:

Source of excitation	Vibration Characteristic
Mass unbalance of rotor	Rotational frequency
Fluid pressure pulsations	Vane passage frequency
Electrostriction of magnetic circuits	Twice alternating current frequency
Turbulent flow	Broad band spectrum
Bearing noise	Higher harmonics of rotational frequency

Vibration transmission from the motor along a drive shaft can be controlled by flexible shaft couplings or belt drives. (See Fig 4). Vibration transmission along the pipework is best controlled by discontinuities such as flexible links introduced into the system or by utilising the natural flexibility of the fluid lines whereby the pipework is "floated" on resilient mounts (having static deflections comparable with the main isolators of the energy source) for a distance of:

"Floated" pipe length	Closest conditioned area
50 pipe diameters	average surroundings (NR 30)
100 pipe diameters	high quality surroundings (NR 25)
200 pipe diameters	critical surroundings (NR 20)

The "floated" pipe run should include bends in two mutually perpendicular directions in order to give three degrees of freedom of movement, with approximately equal distances between successive elbows or bends, (see Fig 5).

Reciprocating sources of pipe-borne vibration most commonly fall into the category of either compressors or internal combustion engines. Vibration of either type of machine arises as periodic variations in gas pressure at the piston and inertia forces at moving parts. Certain multi-cylinder machines may be substantially balanced but one-or two-cylinder machines are inherently unbalanced. In every case gas pressure on the piston reacts at the chassis of the machine to form a force couple which tends to rotate the unit about an axis parallel to the crankshaft. The periodic frequency of the vibration is a function of the speed of rotation of the engine, the number of cylinders and the nature of the operating cycle. Vibration can also arise at unbalanced rotating and reciprocating members, this force acting in planes perpendicular to the crankshaft with components parallel to and perpendicular to the direction of piston motion. Finally a third component of machine vibration will take the form of a reaction couple resulting from variations in torque delivered to or from the machine.

For most reciprocating machines the speeds of rotation are often low and the imbalance forces large. It is often impracticable to isolate the fundamental frequencies since the required deflections at any anti-vibration mounting would tend to cause the machine to be inherently unstable. It is therefore normal practice to install relatively stiff isolators in order to decouple audible frequencies only. The control of pipe-borne vibration originating at reciprocating machines follows the same principles as that adopted for pumps and other rotary sources. For both types of source, it should be noted that vibrational wave transmission along a fluid-filled tube involves not only translational motion of the casing, but also dilatational motions, (see Fig 6). For the latter case the contained liquid is an integral part of the wave transmission medium and a discontinuity in the pipe wall will not provide a very high isolation efficiency for fluid-borne pulsations. In such cases, it may be necessary to include additional proprietary dampers or attenuators to control the periodic motion of the fluid.

In most piped services it is common practice to introduce vibration-isolating flexible rubber or plastic elements in the form of hose or bellows connectors, but metal elements may require to be used where excessive heat or chemical attack is likely. In certain cases it may be necessary, to ensure attachment of a flexible connector at a point of minimum motion, to extend existing connection points by rigid sections to the location of an axis of rocking motion (see Fig 7). For preference, rather than using a long length of flexible pipe work, two shorter units should be connected by a right angle bend or elbow, as illustrated by Fig 8. The free end of a vibration-isolated pipe should be attached to a massive structure (which may - if near a noise-critical area - also require to be isolated from the building structure).

In general flexible pipe couplings take the form of either a bellows unit (fabricated from rubber, plastic or convoluted metal) or a length of hose (also fabricated from either a compliant extrusion or convoluted metal) which are generally re-inforced by an external braid.

3.1 Bellows Connectors

Short metal or rubber bellows are not normally advised for effective vibration isolation unless all the necessary conditions are optimised. Bellows tend to extend under pressure, thereby increasing in stiffness and reducing in isolation efficiency. At the same time they may exert an uncompensated force on the machine. In certain cases extension of bellows units may be countered by vibration-isolated tie rods or by articulation of two units (see Figs 9, 10 and 11).

Metal bellows can occasionally fail in fatigue under the cyclic vibration stresses especially if the excitation frequency is close to the natural frequency of the bellows. If this condition is likely to be encountered, bellows units should be designed to have a natural frequency different from that of the machine. In other words, either the spring rate or the mass of the bellows should be selected to provide the appropriate natural frequency using the formula

$$f_n = \frac{1}{2\pi} \cdot \sqrt{\frac{Kg}{W}} \quad Hz$$

where K = spring rate,
W = mg = weight of bellows.

g = acceleration of gravity.

3.2 Low Pressure Flexible Hose and Electrical Conduit

Where small acceleration amplitudes are to be controlled, short lengths of straight hose may be introduced into pipework, as illustrated by Figure 8. If very large amplitudes of vibration are likely to be encountered, or if the hose has a very low stiffness, it may be used in a looped form provided that the vibrational movement is distributed over its length (see Fig 12). This requirement generally applies to any electrical wiring or flexible conduit carrying power supplies to a prime mover.

3.3 High Pressure Flexible Hose

Flexible pressure tubing generally takes the form of either a concertina flexible tube having spiral or angular corrugations, or reinforced rubber hose (Fig 13). Both are normally enclosed in a metal braid which assists in preventing elongation, controls bursting and helps to damp any free oscillation. Corrugated flexible hoses are restricted in their applications as they cannot accommodate bends and any twisting motion tends to stiffen the link.

Where used to control vibration transmission it is necessary to ensure that flexure takes place in one plane only. Hence it is normal to use two mutually perpendicular links, ensuring that they are installed without any in-built extension, compression or torsion. In each case the axis of the tube should be perpendicular to the vibratory motion at the point of attachment. If used as a connector to a machine in a pressurised circuit which is supported on anti-vibration mountings it will be necessary to make allowances for the added thrust that may be exerted on this machine due to pressure in the couplings. The force so exerted will be equal to the internal pressure multiplied by the pipe cross sectional area (see Fig 9). As an alternative to the balanced pressure unit displayed in Fig 9, it is possible to use a resilient mounting having a load rating equal to that of the applied thrust located under the pedestal supporting the vibration-decoupled pipework.

The required length of metal hose necessary to provide effective isolation will depend on the diameter of the pipework and the vibrational motion of the machine. Table 3 displays the approximate lengths for machines having 3mm movement at 1500 cycles per minute (25 Hz).

Table 3.

Nominal Internal Diameter mm.	"Stiff" hose, length mm.	"Flexible"hose, length mm.
5	160	100
10	190	115
15	210	125
20	230	150
25	250	180
30	280	190
40	330	200
50	380	240
75	500	280
100	600	330

4. Conclusions

As noted previously, pipes and ducts can be decoupled from any support structure at their points of attachment rather than at flexible connectors. This method is preferred in critical locations where the alternative use of pipe couplings and the like may be of an unknown or variable effect. As an example of a critical location we may consider a specialist "floated" enclosure taking the form of a "box within a box" often required for containment of the high amplitude sound fields associated with mechanical plant enclosures. Here it is essential to exclude all vibration transfer to the support structure and hence all pipes and associated services must be supported by anti-vibration hangers, pedestals and the like. Figs 14, 15 and 16 illustrate ways in which this objective can be achieved. Where isolation of acoustic frequencies only is required resilient pipe clamps or support pads can be approved. At wall penetrations no direct contact should be made between the pipework and either of the two leaves. Ideally, the best arrangement would be to allow a clear aperature to surround the pipe but this would give rise to problems of sound or air leakage and makes it necessary to ensure that the hole is plugged by an impermeable seal. Consequently such seals must be resilient and normally take the form of a fibre in-fill capped with a cold-setting rubber compound. It is difficult to achieve this result with conventional low pressure ductwork and it is preferable to grout ducts directly into the wall or slab and decouple from duct borne vibration on either side by means of canvas connectors. High pressure circular ductwork may be treated more like conventional pipework and the appropriate means of support and wall penetration selected.

In conclusion, it must be borne in mind that the control of noise in piped services in buildings is still a relatively unsophisticated procedure, but the general situation is moving towards a more scientific and practical approach. International standards and recommended practices are being put forward for the rating of noise generation by elements of piped systems, and much original research and development remains to be undertaken by both manufacturers and users to determine the methods by which pipe-borne vibration is generated and the best methods for its control.

Duct borne noise and vibration is better understood but with the current trend to ever increasing duct velocities and pressures, the likelihood of this pressure-vessel type of ductwork acting as a vibration transfer path becomes more and more real. Where it is likely that vibration transfer from coupled services will be encountered, a rigorous and systematic approach to the isolation of this vibration must be adopted. It is a good analogy to liken the problems of vibration isolation to that of high-voltage electrical insulation. It only requires one poorly-designed link or contact point to be overlooked to create a flashover or short circuit which can completely nullify the effect of all other isolators included in the system. Consequently any isolated system must be carefully designed from the outset ensuring that adequate provision has been made in the dimensioning and location of the plant to include connectors and isolators which, in general, would be virtually impossible to install as remedial fittings for a system which has failed to meet the design criteria set down for it.

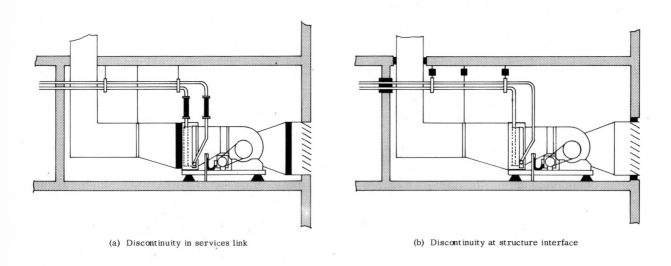

(a) Discontinuity in services link   (b) Discontinuity at structure interface

Fig 1   Vibration isolation of mechanical equipment

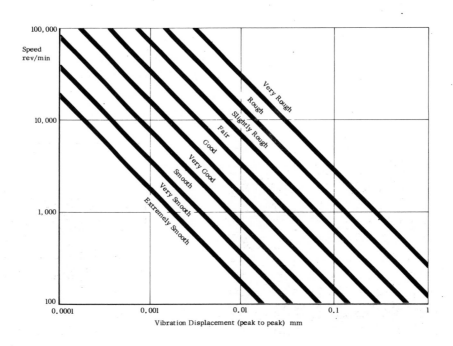

Fig 2   Standards of rotor balance

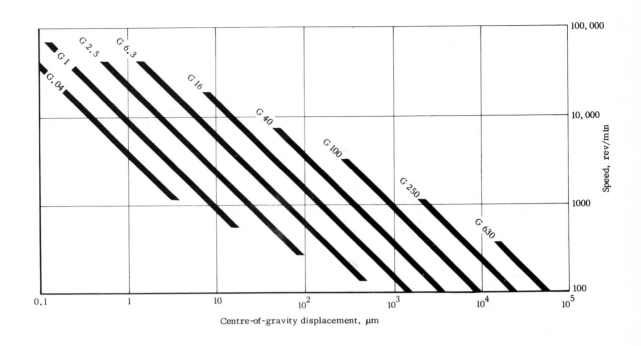

Fig 3   Maximum residual specific unbalance corresponding to various balancing grades, G

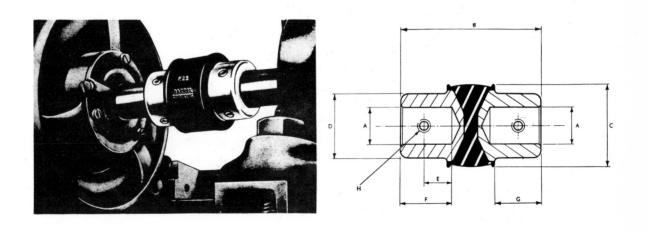

Fig 4   Flexible shaft coupling

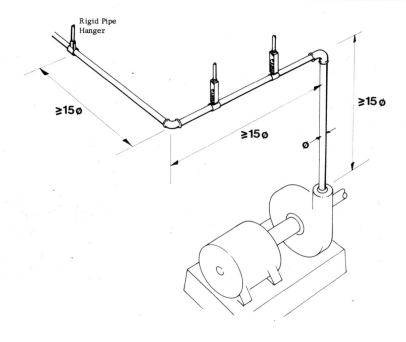

Fig 5   Vibration isolating pipe suspension

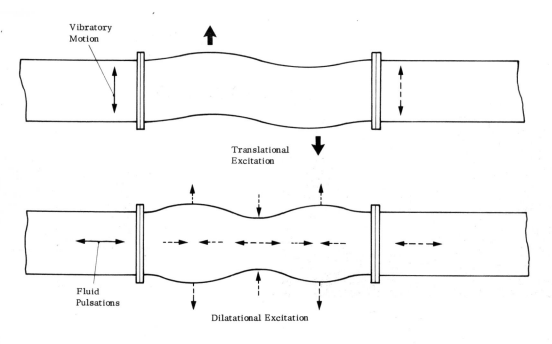

Fig 6   Vibration transfer along flexible link

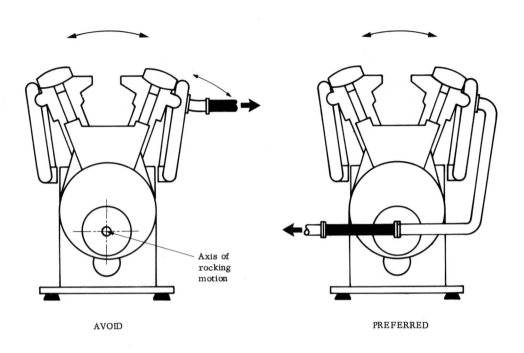

AVOID PREFERRED

Fig 7   Pipe connection to rocking machines

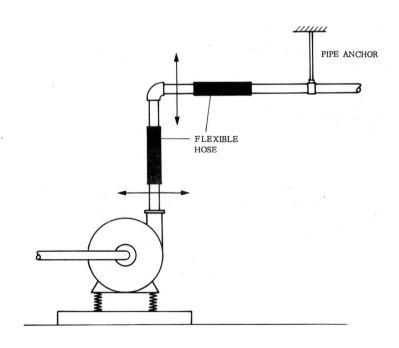

Fig 8   Installation of flexible hose

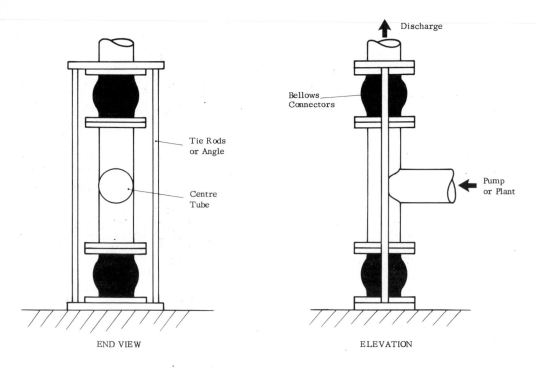

END VIEW   ELEVATION

Fig 9   Compensated bellows connections

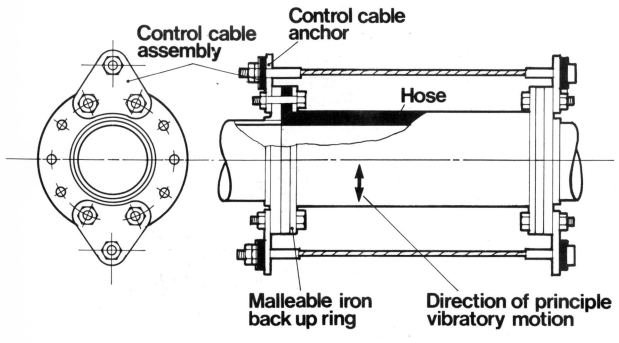

Fig 10a   Flexible hose pipe connector

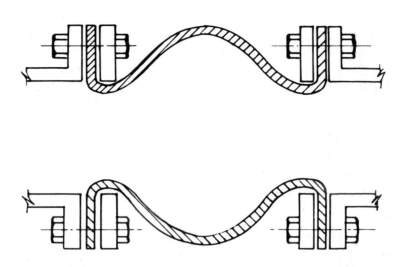

Fig 10b   Rubber expansion joint

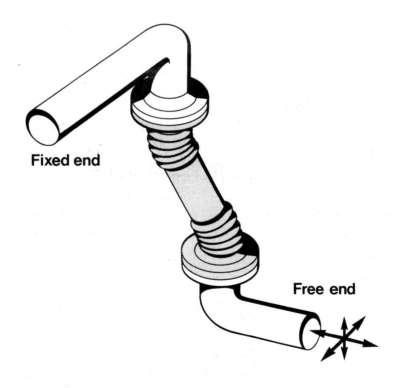

Fig 11   Articulated bellows connectors

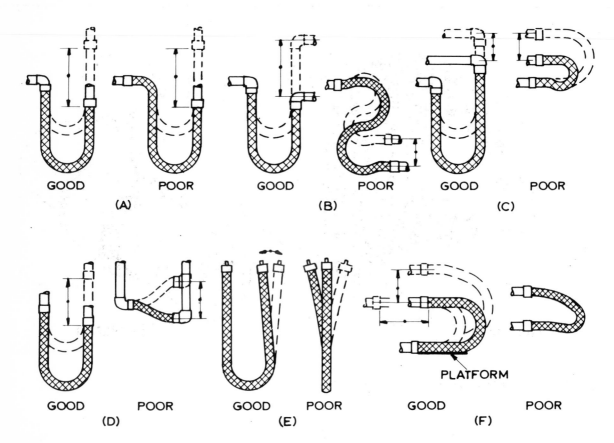

GOOD · POOR (A) · GOOD · POOR (B) · GOOD · POOR (C)

GOOD · POOR (D) · GOOD · POOR (E) · GOOD · POOR (F)

PLATFORM

Fig 12   Optimum arrangement for flexible pipe or cable connectors

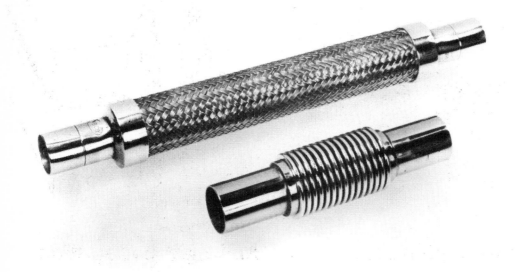

Fig 13   Flexible metallic hose

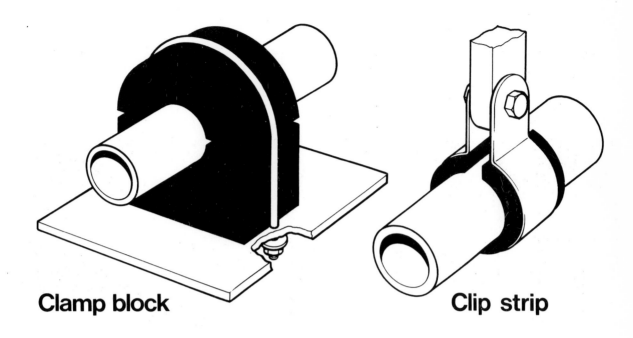

Clamp block   **Clip strip**

Fig 14   Vibration isolated pipe clamps

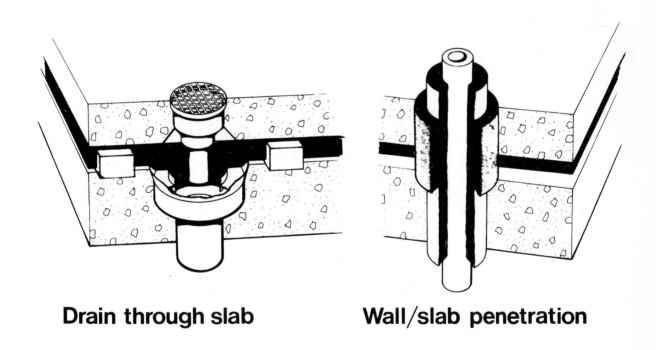

Drain through slab   **Wall/slab penetration**

Fig 15   Penetration of double leaf partitions by services

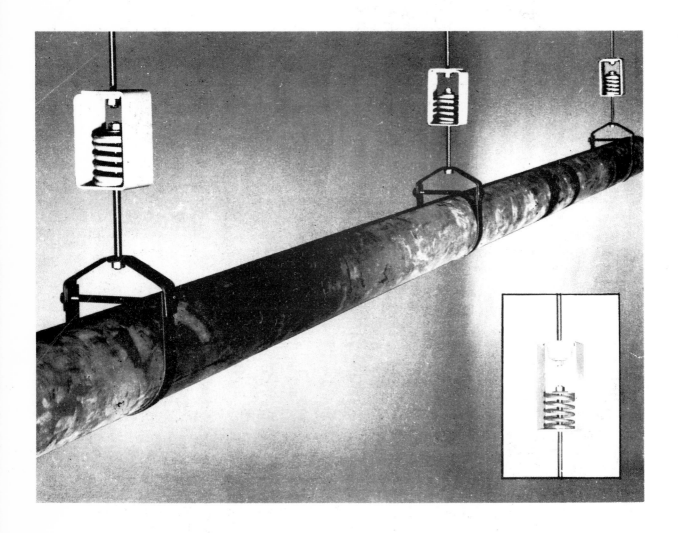

Fig 16   Vibration isolating pipe hangers

Chapter 7
Special Cases

Coupled Modes

Standard selection charts normally consider only one mode of vibration, along the axis of the isolator. Some isolators are actually designed to prevent or limit movement in any other direction. However, these also give little or no isolation in any other direction and are, therefore, limited in use. In most cases this simple approach is adequate, but for extreme cases where plant is very tall for its width, or if large sideways forces are present, a more complicated approach is necessary. In most cases it is merely necessary to be aware of the complications so as to be able to recognise them if they occur.

There are six possible modes of vibration of any free body. There are three translatory modes along three normal axes, Fig. 1, and three rotational modes around these three axes, Fig. 2. Modes of vibration may become coupled, i.e. movement in one mode may cause movement in other modes. Whether this occurs depends on the location and stiffness of mounts and the distribution of the mass of the supported body.

A simple test to establish whether coupling exists is to consider, or apply, a static force directed through the centre of gravity of the system, and examine the resultant motion of the body. If the body moves in two or more modes, these modes will be coupled, e.g. if a force is applied as shown in Fig. 3, the motion of the body will be as shown. This motion has both horizontal translatory motion along the direction of the force, and rotational motion round the z axis. These two modes are, therefore, coupled. This would also be true for horizontal translatory motion along the z axis and rotational motion round the x axis. However, a force applied vertically will cause movement only along the y axis; this mode is, therefore, decoupled.

For the example given above, there will be two resonant frequencies for each pair of coupled modes. If the system is exactly symmetrical, the frequencies will coincide, such that there will be two resonant frequencies for all the coupled modes. There will also be the natural resonant frequencies of the vertical translatory mode and the rotational mode around the y axis.

The resonant frequencies may be calculated for this relatively simple system from:

$$\frac{f_{nc}}{f_{ny}} \cdot \frac{\rho}{\delta_x} = \frac{1}{\sqrt{2}} \sqrt{\frac{\rho^2}{\delta_x^2}\left(1 + \frac{\delta_y^2}{\rho^2}\right) + 1 \ {}^{+}_{-} \ \sqrt{\left[\frac{\rho^2}{\delta_x^2}\left(1 + \frac{\delta_y^2}{\rho^2}\right) + 1\right]^2 - \frac{4\rho^2}{\delta_x^2}}}$$

where f_{nc} = natural frequency in a coupled rotational and horizontal translatory mode

f_{ny} = natural frequency in the vertical uncoupled mode

ρ = radius of gyration of the system

η = ratio of horizontal stiffness to vertical stiffness of the isolators = $\dfrac{Kx}{Ky}$

δ_x & δ_y as in Fig. 4.

The value of f_{ny} can be calculated from the formula used for one degree of freedom systems, i.e.

$$f_{ny} = \frac{1}{2\pi} \sqrt{\frac{Kyg}{W}\left[1 - \left(\frac{b}{b_c}\right)^2\right]}$$

b = damping coefficient and b_c = critical damping coefficient

Where damping is small and can thus be ignored

$$f_n = 15.8 \sqrt{\frac{1}{d}}$$

where d = static deflection of mount (mm)

i.e. the natural frequency is dependent only on deflection. It is, therefore, important to have the same deflection on each mount and one vertical resonant frequency. Equation (1) contains the factor η ; it is obviously desirable that this has a constant value for all the isolators.

Fig. 5 shows Equation (1) in graphical form. It plots $\dfrac{\rho}{\delta_x} \cdot \dfrac{f_{nc}}{f_{ny}}$ against

$\dfrac{\rho}{\delta_x} \cdot \dfrac{Kx}{Ky}$ with varying values of $\dfrac{\delta_y}{\rho}$. There are two plots for each value of

$\dfrac{\delta_y}{\rho}$, giving the two resonant frequencies associated with the two coupled modes

in this example. The values of $\dfrac{\delta_y}{\rho} = 0$, i.e. $\delta_y = 0$ are for the case where the

isolator mounting points are in the same plane as the centre of gravity and the modes are, therefore, decoupled, the horizontal line corresponds to the rotational mode and the inclined line to the horizontal translational mode.

This formula may be further simplified by taking the case where the isolators are attached at the four bottom corners of the body and the body is considered to be cuboid in form with even mass distribution. The results for this situation are shown graphically in Fig. 6. This shows that both the coupled resonant frequencies tend to become a minimum for low values of $\dfrac{Kx}{Ky}$ This is obviously desirable; however, low values of the horizontal stiffness (Kx) can lead to problems of horizontal stability. This implies that it may become necessary to add mass in the form of an inertia block. This will raise the vertical stiffness and hence the horizontal stiffness may also be raised whilst maintaining the low value of the ratio η

The second trend shown by Fig. 6 is that as the ratio of height to width of the body increases, the lower of the two resonant frequencies drops. The behaviour of the higher of the two frequencies, however, depends again on the stiffness ratio (η) and as above it is desirable to have as low a value of this ratio as possible.

Obviously more complicated systems may be analysed; however, the number of variables etc. necessitates the use of a computer.

Isolation of Sensitive Equipment

It is sometimes necessary to isolate very sensitive equipment from the building structure as vibration from many sources within and outside the building will always be present. Typical equipment is: Electron Microscopes, Lasers, and certain electronic equipment. The exact criteria for isolation of such equipment should be obtained from the manufacturer. However, it is often left that the best possible isolation should be provided. This requires as low a natural frequency as possible. This may be achieved with springs, but for example a resonant frequency of 1.6 Hz requires a static deflection of approximately 100 mm and 1 Hz requires 250 mm. Such high deflections will present installation and levelling problems; also the total length of the spring will be very high. We must, therefore, turn to air mounts. There are a number of types of air mounts ranging from the simplest type which is pumped up to the required pressure after installation, to the servo-level controlled mount with extra chambers to provide the correct damping.

Resonant frequencies can be in the range from approximately 1.2 Hz to 3 Hz. Servo level mechanics can be used to correct the level of the isolated body to very low deflections despite high variations in load, and as the static deflection remains constant so does the resonant frequency. Some commercial mounts claim to be highly damped near resonance but with low damping above this. Also, theoretically, the high frequency transmissibility of an air mount is inversely proportional to the square of the forcing frequency (ω^2) whereas a spring is inversely proportional to ω, i.e. the high frequency isolation is better than that of a spring. The ratio of horizontal to vertical stiffness can be made low, e.g. 0.5, thus offering good isolation in all modes.

The gain of the servo mechanism must be limited otherwise overcorrection and an unstable condition of self induced oscillation may occur. This implies that although the servo mechanism may be used to return the mount to a pre-set level, this is not an instantaneous process and will, therefore, not prevent large transient forces causing high deflections. This again depends on the mass of the body and it is usually necessary to use large inertia blocks to control this motion. These may be very large, e.g. in a recent case where it was required that the floor of a small laboratory be isolated, an inertia block of 21 tons was used. It is usually desirable that all the modes should be decoupled, this may be achieved where a large inertia block is used by making the block T shaped. The mounts should be positioned under the top bar of the T section, the size of the bottom section of the T is then selected such that the centre of gravity of the total isolated body is in the same horizontal plane as the top of the mounts and such that the mounts are symmetrically placed round the two vertical planes through the centre of gravity. It will be necessary to space the isolators reasonably far apart to reduce deflection in a rocking mode due to any forces applied directly to the isolated body. However, the natural frequency of a decoupled rotational mode for a simple system as in Fig. 8, is given by

$$f_n = \frac{1}{2\pi} \frac{\delta}{\rho} \sqrt{\frac{Kg}{W}}$$

where f_n = resonant frequency

δ = as in Fig. 8

ρ = radius of gyration (given by $I = m\rho^2$
where I = moment of inertia
m = mass of body)

K = sum of stiffness as in Fig. 8

W = weight of body

This shows that the natural frequency of the system may be decreased by

(1) Increasing the radius of gyration
(2) Decreasing the stiffness
(3) Increasing the weight
(4) Decreasing d.

This last point shows that if the mounts are moved very far apart, the natural frequency in this mode increases and, therefore, the transmissibility increases. A balance must, therefore, be struck between the requirements for as low a natural frequency as possible and for control of rocking motion due to forces applied directly to the body.

Impact Isolation

Previously the aim in selecting isolators has been to have the resonant frequency well below the forcing frequency such that isolation is provided all the way up the frequency range whilst avoiding resonance. Obviously, machines will pass through resonance on run up and run down. As has been discussed earlier, this may be controlled by use of extra mass. However, it is usually a relatively minor problem as the out of balance forces are usually very small at the low frequencies where resonance is encountered. These principles doe not apply for impact isolation. In this case the system must be designed to absorb the energy of each impact before the next impact occurs. If this is not the case, the resultant transmission and deflection can build up and become very large. The time taken for an isolator system to absorb energy is dependent on damping, but it is common practice to assume for a fairly lightly damped system that 6 cycles of free oscillation should be permitted before the next impact, i.e. the natural frequency of the isolators should be 6 times the frequency of impact. In addition, if the resonant frequency of the system were lower than the impact frequency, then at some point during run up and run down, the impact frequency will coincide with the resonant frequency. In this case the force could be as high as during normal running when this point is reached and could cause severe problems. See Fig. 7.

As was mentioned in the first chapter, the displacement is stiffness controlled below resonance and becomes mass controlled above resonance. This applies in impact isolation and, therefore, it is again desirable that the isolated body has a high mass. This implies that for a given deflection (and therefore natural frequency) the stiffness of the isolators is also high.

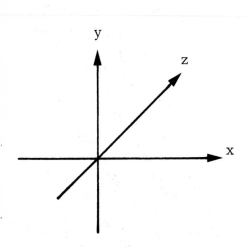

Fig 1

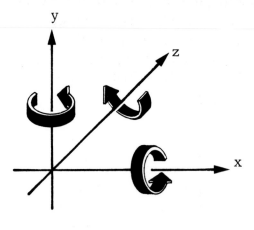

Fig 2

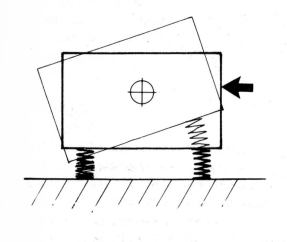

Fig 3

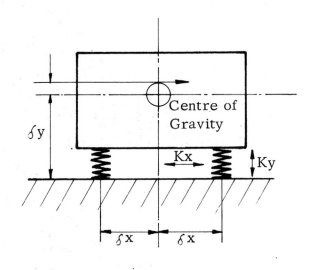

Fig 4   After Harris

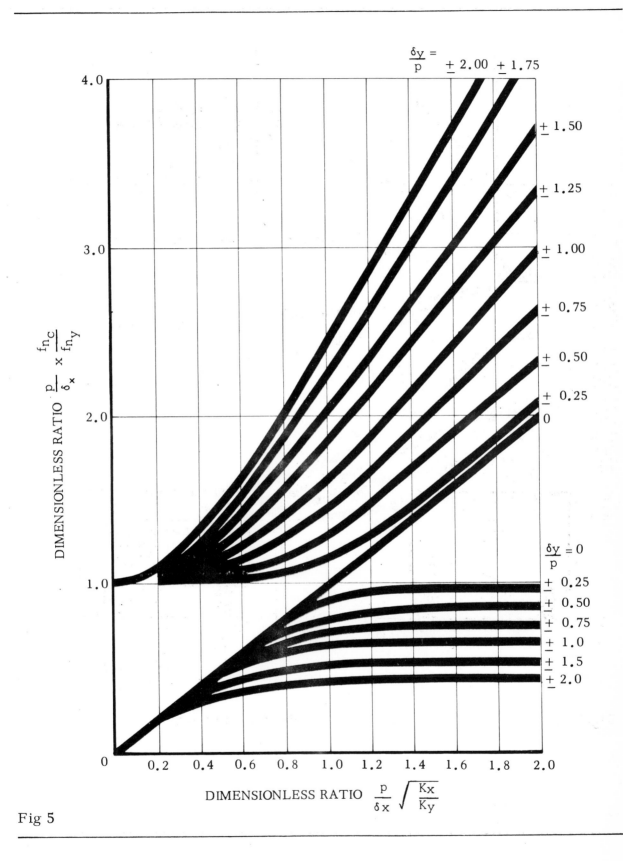

Fig 5

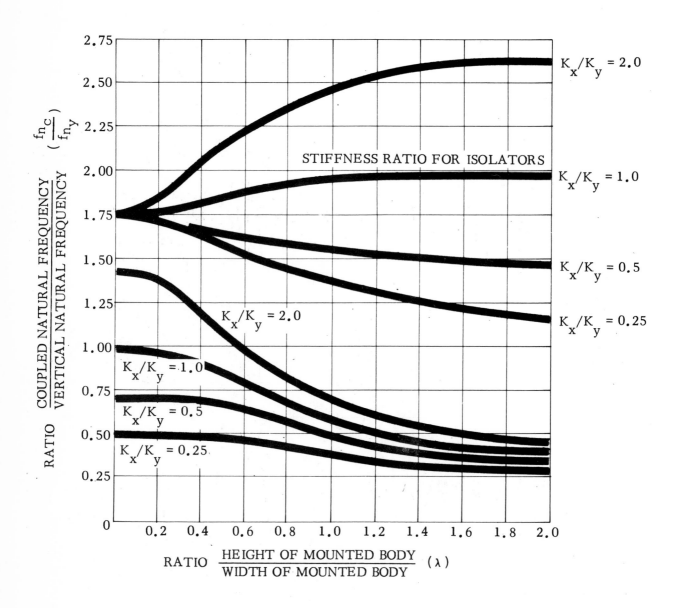

Fig 6

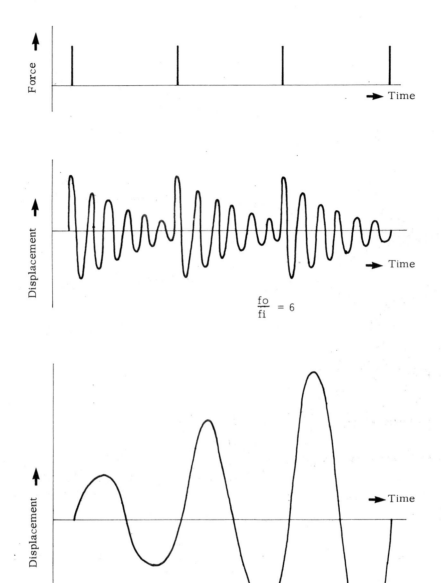

Fig 7

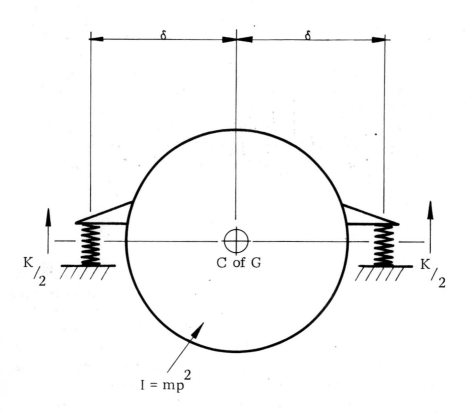

$$I = mp^2$$

Fig 8

Chapter 8
Vibration Measurement ~
Techniques and Instrumentation

Introduction

By vibration, we mean a continuing or steady state periodic motion. This
motion may be simple harmonic like the movement of a pendulum, or it
may be much more complex like the ride experienced in an automobile.
Although sound is transmitted through air as a result of vibrations of the
air molecules, the word vibration is generally associated with structures
and classed as solid borne or mechanical vibration. If these phenomena
are to be reliably measured, it is first necessary for them to be correctly
described in terms of certain characteristic parameters. Having
successfully identified these parameters, it is then possible to devise ways
by which they can be recorded. The purpose of this chapter is first to
discuss how vibratory motions can be described and then to deal with the
instruments and techniques available for their measurement.

Types of Vibration

The most direct way of describing a vibration is by relating the instantaneous
value of its displacement, velocity or acceleration to a reference measuring
point as a function of time. This can be done either in graphical form as
shown in Figs. 1, 2, or in a mathematical form, Fig. 3. The vibration is
said to be periodic when the pattern of motion repeats itself after a length of
time known as the period T. In the case of the single harmonic motion, the
period is inversely related to the frequency of the vibratory motion. All
other more complex forms of periodic motion can be synthesised as a
collection of these sine waves of different frequencies and amplitudes.

Non periodic vibrations are also encountered. Although it is not the intention
to deal with how they may be measured here, it is worth noting in passing the
chief differences between the periodic and non periodic phenomena. The
"random" signal, Fig. 2, is an irregular motion whose pattern never exactly
repeats. Likewise transient signals, often referred to as pulses, perhaps
generated by impacts or explosion, are non repeating; they present very
special measurement and analysis problems.

Periodic phenomena not only have a time domain in which particle
displacement, velocity or acceleration can be plotted against time, they
also have a frequency domain where the vibration, which may be made up of
one or more sine waves, produces a collection of spectral lines when presented
on an amplitude against frequency plot. For example, the square wave can
be represented as a summation of a fundamental frequency sine wave and

all the succeeding sine waves with odd integer multiples of the fundamental frequency.

From Fig. 3 it can be seen that, for a sinusoidal vibratory motion, with displacement plotted against time, there are three distinct values for the amplitude. The greatest displacement is referred to as the peak amplitude A while average amplitude $2A/\pi$ and root mean square amplitude $A/\sqrt{2}$. The latter is used because of its relationship to the energy content of the vibration. It is also shown that for such motion the velocity is proportional to the displacement and frequency of vibration, while the acceleration is proportional to the displacement and the square of the frequency.

Instrumentation for Vibration Measurement

The following applies to the measurement of vibratory phenomena, not shock.

Fig. 4 depicts the four main elements common to most vibration measuring systems. The vibration sensor uses a small amount of energy from the vibration source to generate an electrical signal fed via the preamplifier, to present it in a more usable form for the signal analysis equipment. Here the information content is processed according to particular requirements and finally transferred to some form of signal display or recording arrangement. In Fig. 5 a typical simple measuring system is depicted together with a more complex arrangement for automatic frequency analysis and recording of a source.

Vibration Transducers. Vibration can be defined in terms of displacement, velocity or acceleration; accordingly, there are transducers which can be used to measure each of these directly. Displacement transducers generate an output voltage that is directly proportional to the relative displacement of the pick-up. They usually have poor frequency response and low sensitivity. Velocity transducers generate a voltage that is proportional to the relative velocity of the pick-up, i.e. proportional to both displacement and frequency. They are quite large but have high output voltage with a more uniform frequency response than the displacement transducers. An accelerometer produces an output voltage proportional to the acceleration it experiences.

The schematic drawing, Fig. 6, shows the construction of an accelerometer. A mass is pressed against two piezoelectric discs by means of a stiff spring. The whole arrangement is mounted in a metal housing with a substantial base. When subjected to vibrations, this mass exerts a force on the piezoelectric discs, which is proportional to the acceleration of the mass. The piezoelectric effect of the discs produces a voltage which is proportional to the applied force, and hence the acceleration to which it is exposed.

The voltage produced for a given acceleration will depend upon the transducer's sensitivity. Fig. 7 illustrates the variation of sensitivity both for different instruments and also as a function of frequency. It is desirable to have a constant electrical output, usually given in millivolts per reference acceleration (gravity), over the operating frequency range. The frequency responses are generally very flat up to the 10 kHz to 100 kHz mark, where the instrument's own resonant frequency produces the spike in the response trace.

A further complication is that an accelerometer can have a different frequency response in different directions. Thus a good accelerometer is designed so that its sensitivity perpendicular to its main axis is less than 5% of the main axis sensitivity.

Preamplifiers. These are used either to amplify the relatively low signal from the transducer or to match the transducer impedance to that of the signal analysis equipment. This can be done in two different ways. Voltage preamplifiers, though not uncommon, require care to minimise stray capacitances thus keeping the accelerometer voltage as large as possible. Charge preamplifiers have eliminated the importance of such stray capacitances, and the length of connecting cable from accelerometer is no longer important. Whilst simplifying procedure, this is done only using a more expensive piece of equipment. Fig. 8 shows tables of typical accelerometer and preamplifier performance specification data with comments about their applications and mode of operation.

Integrators are often considered to be a form of preamplifier. It was shown earlier that by differentiation, both velocity and acceleration could be derived from displacement. Conversely, by integration, displacement and velocity can be obtained from acceleration. Therefore, using an accelerometer and the appropriate integrator it is possible to measure all three parameters.

Signal Analysers. As stated earlier, a complex periodic wave form can be considered to be made up of many sine waves of differing frequencies and amplitudes. Signal analysis equipment at one extreme, therefore, can be used to measure the entire frequency content in detail or on the other to look at only the magnitude of the whole signal without considering how energy is distributed as a function of frequency. A signal can be analysed in two ways: constant percentage bandwidth analysis or constant bandwidth analysis. Fig. 9 illustrates the distinction between these two. The type of analysis equipment selected depends on what information processing is required, the speed with which it must be carried out and the way in which the analysed data is presented. Probably the simplest analyser in use is the octave band analyser where the centre frequency of the "window" can be moved by rotating a knob. One third, one tenth and constant bandwidth analysers are also available. For rapid data analysis these are all rather limited since only one bandwidth can be analysed at a time. Thus the narrower the frequency interval of the analyser, the longer the analysis takes. Real time analysers are now being used to overcome the need for instant analysis of signals. By electronic means it is rapidly scanned over its frequency range and the analysed signal presented on a cathode ray display tube.

Recorders. These are devices for the presentation of the analysed signal. Occasionally, as in the real time analyser, they are integrated with the analysis equipment. Cathode ray tubes have the advantage that the analysed wave form can be viewed in its entirety. However, without a camera, they are not as suitable for providing permanent recordings as the graphic level recorder. This usually consists of a servo operated ink pen which reproduces a permanent trace on a strip chart. The strip chart is driven past the pen at a constant speed. Such recorders are valuable for monitoring vibration levels over long periods of time. Magnetic tape recorders are useful for making permanent records of data, particularly in the field, with a view to carrying out analysis of the recorded data at a later stage.

Calibration

It is customary for transducer manufacturers to provide an individual calibration chart with each sensing device. Fig. 10 shows one such chart supplied by Bruel & Kjaer. It gives a plot of frequency response for the accelerometer mounted on a steel exciter of 180 grams with respect to a reference sensitivity. This calibration is better than $\pm 2\%$.

A vibration measuring system consists of more than the transducer and it is often advisable to calibrate the whole system. In order to do this two types of calibrator can be used. The simplest provides a reference vibration level, usually 1 g. at a fixed frequency around 50 Hz. Alternatively calibrators provide larger reference acceleration levels at frequencies which can be swept over a large range. In each case the accelerometer is positioned on the calibrator and the gain of the measuring system adjusted until the indicated value at the recorder agrees with the reference level of the calibrator.

Practical Considerations

Fig. 11 shows six different methods by which a vibration transducer may be mounted on the measurement object. Ideally the mass of the transducer should be negligible compared with that of the vibrating object. This is to minimise its influence upon the nature of the vibrations being investigated. Mounting method one is without doubt the best. Method two is a slight variation to isolate the transducer electrically from the vibrating surface. Similarly method three provides electrical isolation, the transducer being attached via a magnet. The magnetic forces are only strong enough to provide accurate measurements up to 50 g. at best. The method is unsuitable at temperatures above 150°C. Where a threaded stud cannot be used, method four is available. The cement layer should be as thin as possible. Alternatively wax may be used where the surfaces are suitable and at sufficiently low temperatures (method six). The handheld probe is suitable only for approximate diagnostic measurements below 1000 Hz.

To ensure that no stray signals are induced in the accelerometer cable it should be correctly supported as shown in Fig. 11.

The effect of different accelerometer mounting methods upon their frequency response is shown in Fig. 12. These all assume the cable is correctly mounted.

Fig. 13 provides a summary of the procedures that should be observed when carrying out a vibration measurement.

TYPES OF VIBRATIONS (WAVEFORMS), (PERIODIC).

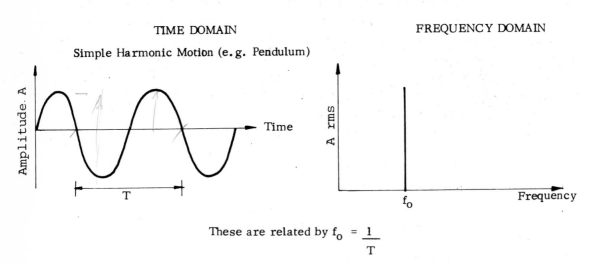

TIME DOMAIN	FREQUENCY DOMAIN

Simple Harmonic Motion (e.g. Pendulum)

These are related by $f_o = \dfrac{1}{T}$

MORE COMPLEX PERIODIC MOTION. (E.G. PISTON ACCELERATION OF I.C. ENGINE.)

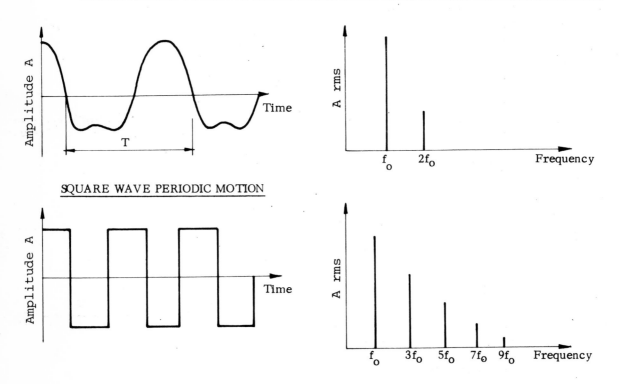

SQUARE WAVE PERIODIC MOTION

Fig 1

TYPES OF VIBRATION, (WAVEFORMS).

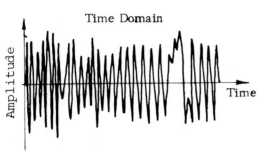

Time Domain

Stationary Random Vibrations

Frequency Domain

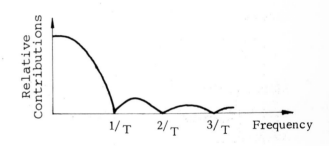

NON PERIODIC VIBRATIONS (TRANSIENTS).

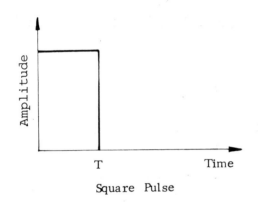

Square Pulse

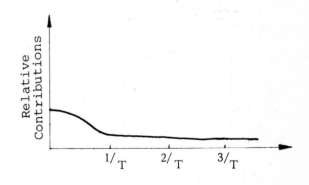

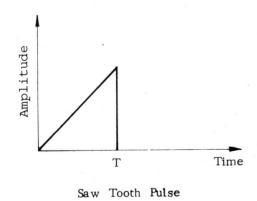

Saw Tooth Pulse

Fig 2

GENERAL DESCRIPTION OF VIBRATION (SHM)

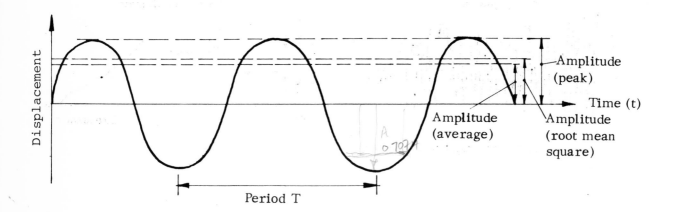

Wave form related by:-

Displacement $x = A \sin \omega t$. ————————— (1)

where $\omega = 2\pi$ (frequency).

t = time

AMPLITUDE (A) = PEAK DISPLACEMENT.

R.M.S. Amplitude = $A/\sqrt{2}$ = .707A.

Average Amplitude = $2A/\pi$ = .636A.

VELOCITY (V) = RATE OF CHANGE OF DISPLACEMENT.

$V = \omega A \cos \omega t$. ————————— (2)

ACCELERATION = RATE OF CHANGE OF VELOCITY.

$a = \omega^2 A \sin \omega t$

or $a = \omega^2 x$ ————————— (3)

i.e Acceleration = ω^2 (Displacement).

NB. These formulae apply only to sinusoidal vibration.

REFERENCE QUANTITIES

VELOCITY 10^{-8} metres/second (10^{-5} mm/s).

ACCELERATION 10^{-5} metres/second^2 (10^{-2} mm/s^2).

Fig 3

TYPES OF VIBRATION, (WAVEFORMS).

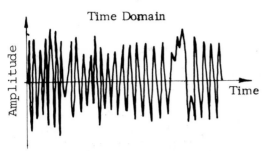

Time Domain

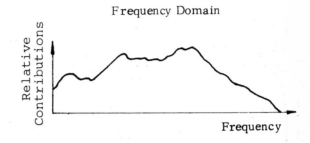

Frequency Domain

Stationary Random Vibrations

NON PERIODIC VIBRATIONS (TRANSIENTS).

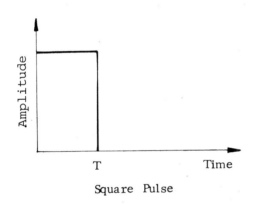

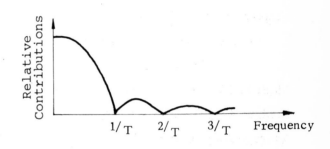

Square Pulse

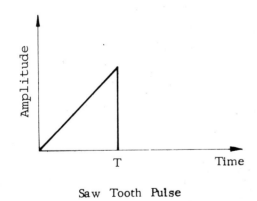

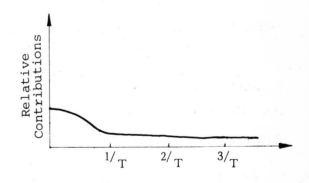

Saw Tooth Pulse

Fig 2

GENERAL DESCRIPTION OF VIBRATION (SHM)

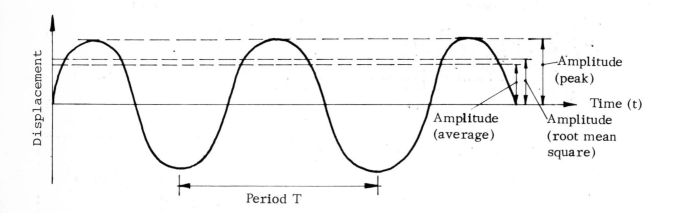

Wave form related by:-

Displacement x = A Sin ωt. ————————— (1)

    where ω = 2 π (frequency).

          t = time

AMPLITUDE (A) = PEAK DISPLACEMENT.

R.M.S. Amplitude = $A/\sqrt{2}$ = .707A.

Average Amplitude = $2 A/\pi$ = .636A.

VELOCITY (V) = RATE OF CHANGE OF DISPLACEMENT.

        V = ωA Cos ωt. ————————— (2)

ACCELERATION = RATE OF CHANGE OF VELOCITY.

        a = ω^2 A Sin ωt

    or a = ω^2 x ————————— (3)

i.e Acceleration = ω^2 (Displacement).

NB. These formulae apply only to sinusoidal vibration.

REFERENCE QUANTITIES

    VELOCITY 10^{-8} metres/second (10^{-5} mm/s).

    ACCELERATION 10^{-5} metres/second^2 (10^{-2} mm/s^2).

Fig 3

VIBRATION MEASURING SYSTEMS, (SCHEMATIC)

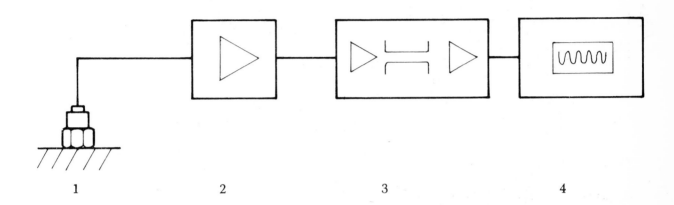

1. VIBRATION TRANSDUCER. Converts mechanical vibrations of surface into an analogous electrical signal.

2. PREAMPLIFIER. This matches the higher electrical impedance of the vibration transducer to the analysis equipment. It can also amplify a weak transducer signal.

3. AMPLIFIER/ANALYSIS NETWORKS. The electrical signal is further amplified and processed according to the information required.

4. RECORDER/READ OUT DISPLAY. The analysed result is presented in digital or analogue form.

Fig 4

Simplest form of measuring system for octave band analysis of vibration signal (after B & K)

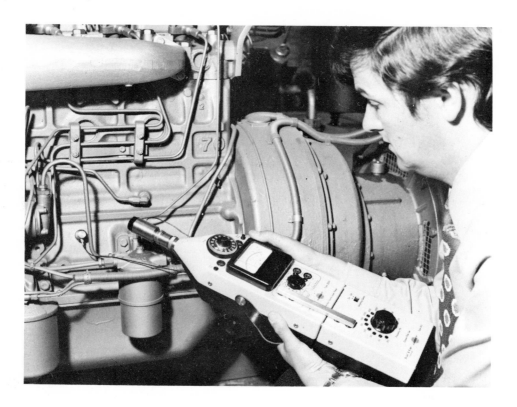

Instrumentation for automatic frequency analysis of a vibration signal (after B & K)

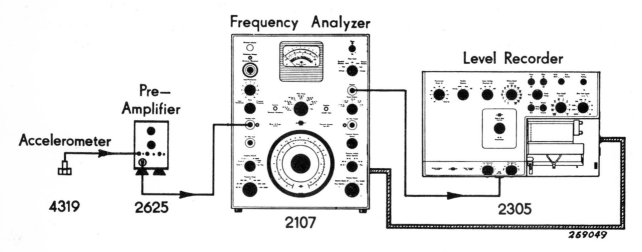

Fig 5   Typical automatic frequency analysis instrumentation

Construction (Schematic)

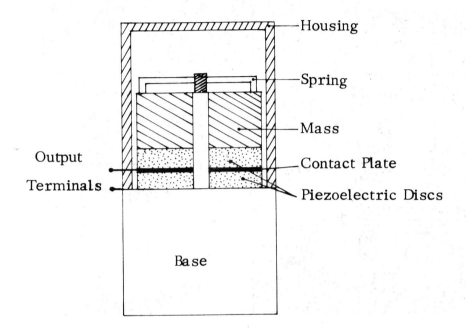

Output voltage at terminals is proportional to acceleration to which the transducer is subjected. It results from the action of the mass held against the piezoelectric discs by a stiff spring.

Fig 6   Accelerometers - construction

Typical frequency response characteristics (main axis)

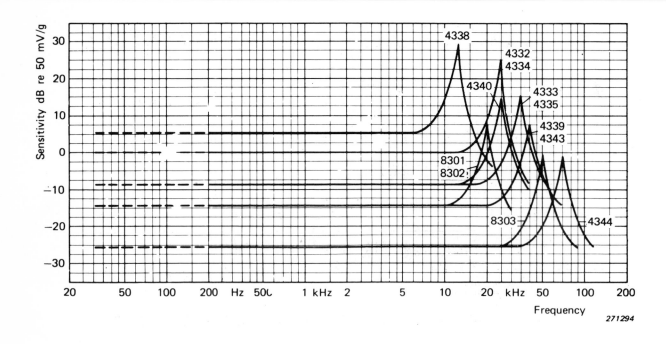

For a good accelerometer, the transverse sensitivity should be less than 5% of the sensitivity in the direction of the main axis

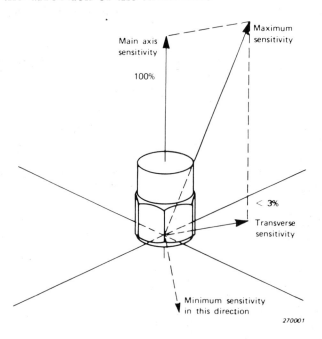

Fig 7   Accelerometer frequency response/sensitivity (after B & K)

Typical accelerometer performance/selection data (after B & K)

Accelero-meter Type	Main Application	Suitable Preamplifier(s)*)	Sensitivity		Mounted Resonance (kHz)	Transverse Sensitivity (%)	Maximum Shock (g)	Weight (Grammes)
			Voltage (mV/g)	Charge (pC/g)				
4332	General Vibration Measurements. Vibration Testing and Control.	1606, 2616, 2622, 2623, 2625, 2204 (+ ZR 0020)	45–65	40–60	30	< 4	7000	30
4334					30			
4333	High Frequency and General Vibration Measurements.		14–24	14–20	40	< 4	10000	13
4335					40			
4338	Building Vibration Measurements. Measurement of Low Level, Low Frequency Vibrations. For Use with Charge Sensitive Electronics.	**2624** (1606, 2616, 2623, 2625, 2204 + ZR 0020)	~ 100	100 ± 2	12	< 3	2000	~ 60
4339	General Vibration Measurements. Vibration Testing and Control. For Use with Voltage Sensitive Electronics.	1606, 2616, 2622, 2623, 2625 (2204 + ZR 0020)	10 ± 0.2	~ 10	45	< 3	10000	16
4340	For Vibration Measurements in Three Mutually Perpendicular Directions.	**2625** (3 × 2616, 3 × 2623, 3 × 1606)	14–24	14–20	23	< 4	500	35
4343	General Vibration Measurements. Vibration Testing and Control. For Use with Charge Sensitive Electronics.	**2624, 2622** (1606 2616, 2623, 2625, 2204 + ZR 0020)	~ 10	10 ± 0.2	50	< 3	10000	16
4344	High Frequency (and High Level) Vibration Measurements. Shock Measurements. Measurements on Light Weight Structures.	1606, 2616, 2623, 2625, 2203 (+ ZR 0020)	2–3	2–3	70	< 5	14000	~ 2

*) All Brüel & Kjær Accelerometers may also be used together with the Microphone Preamplifiers Type 2614, 2615 and 2619.

Typical preamplifier performance/selection data (after B & K)

Preamplifier Type	Features	Powered by	Frequency Range	Input			Output		Dynamic Range
				Resistance	Capacitance	Max. Inp. voltage	Resistance	Max. Outp. voltage	
1606	Contains integrating networks for the measurement of velocity and displacement. Max. amplification: 38 dB (80×). Built-in accelerometer calibrator.	2107, 2112, 2114, 2603, 2606, 2801, 2803, 2804 Mains	At 0 dB amplification 0.2 Hz–100 kHz. At max. amplification 2 Hz–20 kHz	200 MΩ	50 pF	10 V peak (0 dB amp.)	–	20 V peak	120 dB (10⁶)
2616	Battery operated. Battery compartment removeable. Contains overload indicators and built-in signal attenuator (– 40 dB)	6–35 Volts DC- (4–10 mA)	0.5 Hz–500 kHz	> 1200 MΩ	10 pF	1.4 V peak (140 peak)	< 100 Ω	1.4 V peak	90 dB (3×10⁴)
2622	Contains two identical channels. For voltage or charge operation. Digital scaling to accelerometer sensitivity. Built-in low and high frequency cut-off filters. Specifically suited for vibration and shock testing applications	Mains	2, 5, 10 or 50 Hz to 10 kHz	Voltage Operation > 600 MΩ	Voltage Operation 16 pF Charge Operation 0.75 μF	40 mV peak at 2 Hz 20 V peak from 40 Hz to 10 kHz	< 200 Ω	–	~ 100 dB (10⁵)
2623	For impedance transformation. Small, lightweight instrument, insensitive to mechanical vibrations. Powered from external DC source	2805 (28 Volts DC)	0.5 Hz–500 kHz	> 2000 MΩ	3.5 pF	~ 10 V peak	40 Ω	~ 10 V peak	~ 86 dB (2×10⁴)
2624	Charge amplifier for the measurement of very low frequency vibrations and shocks. Eliminates influence of cable between accelerometer and pre amplifier. Powered from external DC source.	2805 (28 Volts DC)	0.03 Hz–30 kHz	(> 10 GΩ)	–	–	< 5 Ω	~ 10 V peak	–
2625	Contains integrating networks for the measurement of velocity and displacement. Three suitable inputs with individual adjustments. Powered from DC. Built-in battery compartment. Max. Amplification: 20 dB (10×).	2805 (28 Volts DC)	1 Hz–35 kHz (Acceleration)	> 450 MΩ	14 pF	5 V peak at 0 dB gain	< 50 Ω (from serial no. 277995)	5 V peak	~ 86 dB (2×10⁴)

Fig 8

Constant Bandwidth Analysis.

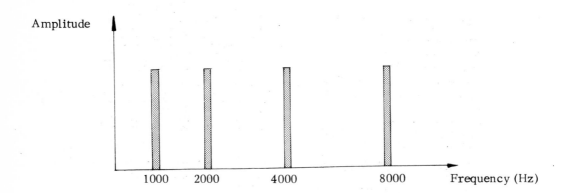

Here the band width of the "window" through which a particular part of the spectrum is observed, is always fixed, independent of the analysis frequency.

Constant Percentage Bandwidth Analysis.

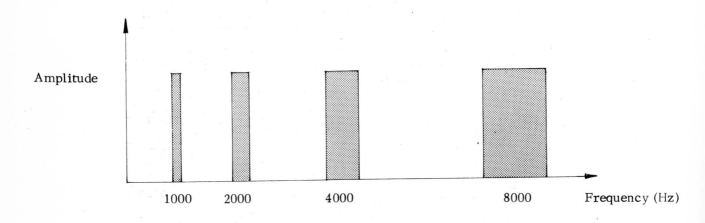

Here the width of the "window" is proportional to the analysis frequency.

Fig 9   Frequency analysis of accelerometer signal

**Calibration Chart for
Accelerometer Type 4332**

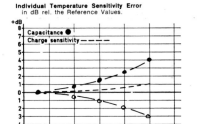

Brüel & Kjær
Denmark

Serial no. ..*415945*

Reference Sensitivity at ...*50*.. Hz at ..*23*... °C
and including

Cable Capacitance of ....*108*.... pF:

Voltage Sensitivity** ......*62,7*... mV/g*

Charge Sensitivity .....*72,7*..... pC/g

Capacitance (including cable) ....*115,9*.. pF

Maximum Transverse Sensitivity at 30 Hz ...*2,7*... %

Weight ...*30,5*...... grams

Undamped Natural Frequency ....*46*..... kHz
For Resonant Frequency mounted on steel exciter
of 180 grams and for Frequency Response relative
to Reference Sensitivity, see attached individual
Frequency Response Curve.

Polarity is positive on the center of the connector for
an acceleration directed from the mounting surface
into the body of the accelerometer.

Resistance minimum 20,000 Megohms at room temperature.

Date ..*26-6-73* Signature ...*M.P.*.....

• 1 g = 980.6 cm sec⁻². $\frac{mV}{g} = \frac{mV_{RMS}}{g_{RMS}} = \frac{mV_{peak}}{g_{peak}}$

** This calibration is traceable to the National
Bureau of Standards Washington D.C.

BC 0063

Individual Temperature Sensitivity Error
in dB rel. the Reference Values.

Capacitance ●
Charge sensitivity — — — —
Voltage sensitivity ○

Physical:

Material: Stainless Steel
Mounting Thread: 10–32 NF
Electrical Connector: Normal coaxial
10–32 thread

Environmental:
Humidity: Sealed
Max. Temperature: 260°C or 500°F
Shock Linearity: 3,000 g typical for 200 μsec half sine
wave pulse or equivalent.
Max. Shock Acceleration: 6,000 g typical
Magnetic Sensitivity (50-400 Hz) < 1 μV/Gauss
Acoustic Sensitivity < 0.2 μV/μbar

For further information see instruction book.

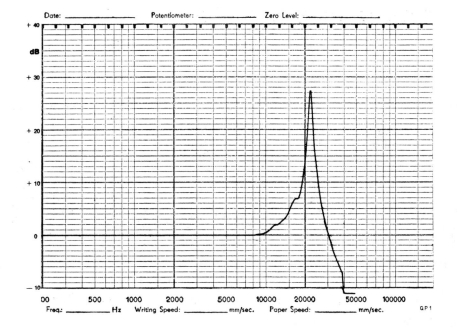

Date: _____ Potentiometer: _____ Zero Level: _____

Freq.: _____ Hz   Writing Speed: _____ mm/sec.   Paper Speed: _____ mm/sec.   QP 1

Fig 10

3.2.2

Accelerometer mounting methods (after B & K)

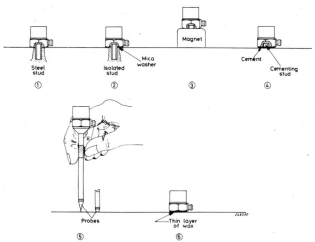

Different methods of mounting of the accelerometers:
1. With steel stud.
2. With isolated stud and mica washer.
3. With permanent magnet.
4. With cementing stud.
5. Handheld with probe.
6. Accelerometer stuck on with wax.

Accelerometer cable must be correctly positioned (after B & K)

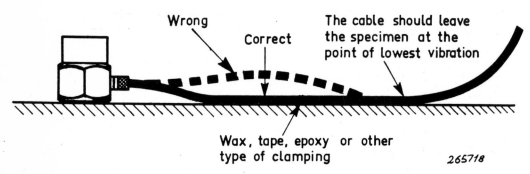

Clamping of the cable to avoid relative movements (cable "whip".)

Fig 11

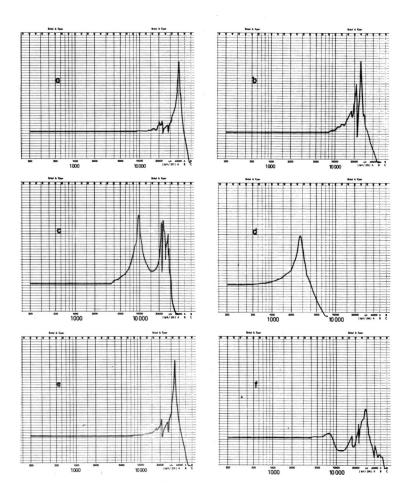

Typical frequency response curves for an accelerometer when mounted according to some of the methods suggested in Fig 11.

a) Mounting by means of steel stud ___ the best
b) Mounting by means of isolated stud and mica washer
c) Mounting by means of permanent magnet
d) Handheld with probe
e) Mounting by means of wax
f) Mounting by means of soft glue (not recommended)

Fig 12   Accelerometer frequency response as a function of mounting condition

1. Select mounting position for vibration transducer.

2. Estimate vibration characteristics and magnitudes at this point.

3. Select transducer accordingly.

4. Identify what information is required from the measurement.

5. Select preamplifiers, signal processors and recording equipment as appropriate.

6. Connect equipment correctly and calibrate.

7. Record instrumentation and connection details.

8. Pick accelerometer mounting method appropriate to 1, 2 and 4.

9. Mount accelerometer as appropriate, measure and record results.

10. Note settings of all instrument control knobs and other relevant information.

Fig 13   Vibration measurements - recommended procedure

Chapter 9
Overall Vibration Control

Introduction

The following notes are intended to summarise the basic contents of all
preceding chapters. They are presented as an 'aide-memoire' to assist
in any designs for vibration control.

Vibration Control Engineering

The evolution of a practical solution to a vibration control problem involves
five basic actions:

Action	Procedure
1. Define the vibration excitation characteristics	Analyse the vibration environment and develop a simplified mathematical model of the excitation characteristics
2. Specify all vibration control performance requirements	Relate to subjective design criteria, vibration tolerance of machinery, stability and performance limits, structural integrity, etc.
3. Select appropriate vibration control methods	Review all applicable techniques and select one or a combination of methods to attain the performance objectives
4. Undertake analytical design of vibration control system	Proceed with theoretical analysis of systems considered under 3 above in order to quantify the static and dynamic performance of the selected system
5. Select necessary materials and/or fittings	Review appropriate products and control systems to meet constraints imposed by available space, loadings, methods of attachment, stability, visual appearance, etc.

The following notes extend and develop some of the factors to be considered in
the execution of the various stages of a design for vibration control.

Task 1

a) Determine whether the objective of the exercise is to isolate the vibration source from a structure (<u>positive</u> or <u>active</u> isolation) or to isolate a vibrating system from a potential recipient (<u>negative</u> or <u>passive</u> isolation). (Active vibration isolation normally involves consideration of both sub-audible (3-30Hz) and audible (30-1000Hz) frequencies. Passive vibration isolation normally involves the range of frequencies sensed by touch only (3-30Hz).)

b) If the vibration arises at rotating machinery, consider its degree of balance, rotational frequencies of all shafts, gear tooth meshing frequencies, out-of-balance forces in reciprocating machinery, compressor output pressure pulsations and all higher harmonics of fundamental frequencies and beat frequencies of all shafts and excitation mechanisms.

Task 2

a) The limits for vibration transferred to buildings will be determined for the audible frequency range by the appropriate background noise criterion (NC, NR or dBA ratings).

b) For sub-audio frequencies, the limits will be given by the Reiher-Meister or Deickmann systems of subjective ratings. Other systems can give advised limits for avoidance of structural damage, fatigue failure, etc., or the maintenance of machine stability, etc.

Task 3

Seven basic methods may be considered for the reduction of vibration amplitudes:

Method	Procedure
a) Reduce dynamic excitation	Improve inertial balancing; improve quality of manufacture; modify or redesign vibration source to reduce vibration acceleration (eg reprofile cams, reduce throw of cranks, reduce oscillating masses, etc.).
b) Increase structural rigidity	The deflection amplitude of a vibrating member may be decreased by stiffening (which will increase its resonant frequency and its overall strength). Check that all fundamental and harmonic resonances of the stiffened structure occur or frequencies outside the range of vibration excitation.

c) Detune resonant frequencies

Design equipment to ensure that resonant frequencies of members and components differ from one another and from predominant excitation frequencies. No natural frequencies of the assembly should coincide.

d) Decouple vibration

Design equipment to ensure that coupled resonators are minimised in number or avoided altogether.

e) Isolate vibration

Interpose a resilient element between the vibration source and the receiver. This is a 'broad band' control technique which can limit both the fundamental frequency and its harmonics. 'Stiffness mechanisms' used as resilient elements can consist of: Metal springs (strip or coil)
Elastomers/rubber
Elastomeric foam
Cork, felt or composite materials
Wire mesh
Stranded mineral fibres
Pneumatic/hydraulic systems
Check that the selected system provides sufficient inherent damping for energy dissipation or add dampers as necessary.

f) Absorb vibration

Attach an energy absorbing mechanism (a combination of a mass and a resilient element to function as an energy sink). This is a narrow-band technique and can only be employed to control excitation at a single frequency. The level of vibration attenuation will depend on the degree of absorber damping and the degree of synchronism of absorber frequency and critical frequency.

g) Damp structural resonances

Apply high energy dissipating coatings for fixtures to reduce the amplitude of vibration of structural resonances. These may take the form of visco-elastic coatings, visco-elastic-damped structural composites, built-up structural assemblies (slip-friction damping at rivets or bolts) or special proprietary dampers (slip-friction, viscous-shear or viscoelastic-shear damping).

Task 4

a) For simple well-defined problems use 'short-form' charts or graphs to quantify isolated performance in terms of their static deflection or resonant frequency.

b) For complex systems, a detailed analysis of the vibration response of all elements may be necessary, involving specialist design services and the use of a powerful computing facility.

c) Quantify the damping performance of isolators or added treatment. Where necessary, establish the degree of compromise between resonant vibration control and high frequency vibration isolation.

Task 5

a) Having established the preferred combination of techniques for vibration control, examine all potential proprietary equipment to determine their conformation with design requirements and their performance limits. Where necessary establish the degree of compromise between the desired and attainable targets.

b) Check that selected equipment or treatments conform to requirements for resistance to fire, freezing, chemical attack, corrosion, high temperature working, etc.

c) Ensure that final design makes provision (where necessary) for levelling the machine, mount compression by shipping bolts, access for servicing and inspection of treatment, etc.

Notes on the Authors

N.A. Grundy, B.Sc., M.Inst.P., MCIBS, MIOA, M.Inst.M.

After graduating from the School of Physical Sciences at the University of Sussex, he went to the Leyland Motor Corporation gaining experience as a graduate apprentice in the automobile industry. He then joined the acoustic group at Imperial College on a scholarship from Sound Attenuators Limited, to study for the M.Phil. Degree. In his research he investigated new methods of sound attenuation in air distribution systems. He moved to Colchester in 1970, becoming involved in laboratory activities handling all aspects of product development and materials testing. He joined SRL as Laboratory Manager in 1973 and was made a director of the company in 1975 and Managing Director in late 1975.

T.J.B. Smith, B.Sc., Ph.D., C.Eng., MIEE, M.Inst.P., FIOA

After graduating from the Department of Electrical Engineering at the University of Leeds, Dr. Smith gained his Ph.D. for a study of combustion noise in the Houldsworth School of Applied Science at the University of Leeds. Following this, he spent three years in the University of California at Los Angeles working in the aerosonics laboratory of the Department of Engineering on a wide variety of problems associated with noise and fluid flow. On returning to the United Kingdom, he joined the research department of the BBC at Kingswood Warren where he was involved in the development of a wide range of acoustical applications to broadcasting, including studio design, systems for artificial reverberation and the development of lightweight sound insulating partitions. Dr. Smith has been a Director of Sound Research Laboratories since 1967. He is a member of the Council of the Institute of Acoustics and a member of the Acoustical Society of America.

J.D. Webb, B.Sc.(Eng), Ph.D., C.Eng., M.I.Mech.E., FIOA

Combined a three year mechanical engineering degree at Imperial College, London, with a student apprenticeship at the National Gas Turbine Establishment, Pyestock. After graduation, he studied flow in axial flow compressors for a Ph.D. also at Imperial College. In 1960 he joined the English Electrical Company in their Mechanical Engineering Laboratories where he worked on general problems of air flow and fluid dynamics. In 1963 he helped form a new section set up to deal with problems of high intensity acoustics which were arising in large power plants such as the gas circuits of nuclear power stations and large steam turbines. He later became involved in research on general acoustics problems, as head of the dynamic analysis section of the Mechanical Engineering Laboratories. In 1968 he joined Sound Research Laboratories Limited at Colchester as a Consultant and was made a Director in 1971.

G.J. Cole, MCIBS

Gained experience in mechanical services installations during an eleven year term with a leading London H & V contractor. During this period which included the diploma course at the National College, he became responsible for the design and installation supervision of all forms of heating, ventilating and air conditioning contracts. His responsibilities at Sound Research Laboratories are mainly noise control and planning associated with mechanical services installations and building acoustics. He was made a Director in 1975.

J.R. Cowell, B.Sc., B.Arch., M.Sc., RIBA, MIOA

Is a fully qualified architect who joined the permanent staff of SRL in 1970 having worked in conjunction with SRL staff on research at Strathclyde University. The particular field of study was architectural and building science, together with research into the performance of portable acoustic screens in open spaces. Prior to this period, he was employed in architecture, quantity surveying, and on building sites including a year in an architect's office in Toronto, Canada. He handles mainly architectural acoustics, materials testing, acoustics and music, acoustic model studies, working details and the building context of consultancy work.

T.P.C. Bramer, B.Sc.(Eng), C.Eng., MIEE

Obtained his honours degree in electrical engineering from London University after specialising in the electronics and telecommunications fields. After working in electronics applications for Thorn-AEI Ltd., and teaching in a public school, he joined the acoustics section of Standard Telecommunications Laboratories, Harlow, where he worked for five years on the electro-acoustics of telecommunications systems. His primary consulting responsibilities at SRL now include electro-acoustics and all aspects of industrial and environment acoustics, including noise control, hearing damage risk and the legal aspects of noise.

D.R. Winterbottom, B.Sc., MIOA

Graduated in Physics at Exeter University in 1971. After working for a short time in a sound recording studio, where he became interested in acoustics, he joined SRL in December 1971. Initially he worked as a Consultant in the Industrial and Environmental Division, progressing on to the Laboratory where he pioneered Vehicle Noise Control. This has now lead to the formation of the Vehicle and Machinery Noise Control Division. He is an Associate Member of the Institute of Physics and a Member of the Institute of Acoustics.

A.T. Fry, B.Sc., ARCS, M.Inst.P.

Director of Sound Attenuators Limited

After graduating from the Physics Department of Imperial College, London, he served a short apprenticeship with Marconi at Chelmsford. During this time he was involved with the development of the high frequency radio transmitters. On leaving Marconi he returned to Imperial College to study for a Ph.D. - the specialist field being the isolation of explosive type shocks. In 1963 he took up a lectureship in physics and mathematics at the College of St. Mark and St. John, Chelsea, where he instructed teachers to degree level. He continued his research at Imperial College and also lectured on their specialist acoustic courses, together with those for the M.Sc. degree at Chelsea College of Technology. Two research studies supervised by him at Imperial College were those on the use of acoustic ducts for predicting the transmission loss of materials and the other on acoustic prisms.

Terminology

These definitions are intended to be helpful rather than rigorous

Acceleration   Rate of change of velocity. Measured in m/sec^2 (ft/sec^2) or in terms of the acceleration due to gravity ($g = 9.81$ m/sec^2 or 32.2 ft/sec^2). Thus an acceleration of 20 $m/sec^2 = 2.04$ g. Vibration is normally thought of in displacement terms i.e. physical movement, but it may equally well be described or measured in acceleration terms.

Acceleration due to gravity   The rate at which a free falling object will accelerate (9.81 m/sec^2 or 32.2 ft/sec^2).

Accelerometer   A vibration measuring transducer that produces an electrical output proportional to acceleration. The term is sometimes used loosely to cover any vibration transducer.

Amplitude   Strictly speaking the amplitude is the maximum level of any periodic quantity, but it is usually used as amplitude of displacement i.e. zero-to-peak displacement.

Anti Vibration Mount   See vibration isolator.

Critical Damping   A system with critical damping will not oscillate when displaced from rest and released. It returns to its original position as quickly as possible without overshoot. If it is more than critically damped it returns more slowly to rest.

Critical Speed   The speed of rotation of a shaft which corresponds to a system resonance. More particularly, the speed at which a shaft starts to whip due to an unbalanced centrifugal force.

Coulomb Damping   Damping where the damping force is independent of velocity. Dry friction is a typical example.

Damping   Any influence that extracts energy from a vibrating system.

117

Damping Controlled   A system is said to be damping controlled when vibrating at its natural frequency. Under this condition the effects of mass and stiffness cancel out and only damping controls the amplitude of the vibration.

Damping Ratio   The damping present in a system expressed as a fraction of the critical damping.

Displacement   The distance that a vibrating body moves from its rest position. It may be measured as instantaneous, peak-to-peak, zero-to-peak or root-mean-square values.

Dynamic Absorber or Damper   A subsidiary damped spring-mass system attached to a vibrating component and tuned so that the dynamic absorber vibrates in resonance. In so doing energy is dissipated in the absorber and reduces the vibration of the vibrating component. A common application is the use of a steel rimmed rubber disc on the front of the crank shaft of a car engine to damp out torsional vibration.

Forced Vibration   When vibration results from a periodic disturbing force (the forcing frequency) it is called forced vibration.

Free Vibration   Vibration that occurs when an elastic system is displaced from its rest position and released.

Frequency   The repetition rate of a periodic process. It may be measured in cycles or revolutions per minute, or Hz (i.e. cycles per second).

Hysteresis Damping   Damping due to a non-reversible process, for example, when a rubber block is compressed and released it does not release all of the energy due to compression. If the process is repeated the rubber becomes hot due to the "lost"energy. An example of low hysteresis rubber occurs in the toy "super" balls which bounce more than a normal rubber ball.

Moment of Inertia   The tendency of a body due to its mass and shape to resist changes in angular velocity. The moment of inertia is numerically equal to the mass times the (radius of gyration)^2. The radius of gyration is the radius of an equivalent wheel of the same mass with all its mass concentrated in the rim (like a bicycle wheel). The values of the radius of gyration for standard shapes are tabulated in reference books.

Natural Frequency   A frequency at which a system vibrates when disturbed from rest and then released. An idealised spring-mass system has one natural frequency. Real life systems have many natural frequencies. The more complex the system the more natural frequencies it will have.

Period   The time between successive cycles of a periodic quantity.

Periodic quantity   A periodic quantity is one that repeats itself at equal intervals of time.

Resonances   When a spring mass system is subject to a disturbing force at a natural frequency a very large oscillation will be set up. An increase or decrease in the forcing frequency causes a decrease in the oscillation. This state is known as resonance.

Resonance Frequency   The frequency at which a resonance occurs. For lightly damped systems resonance occurs at the natural frequency. For very large values of damping the resonant frequency may change slightly.

Root Mean Square   Vibration levels are normally measured in terms of the root-mean-square displacement, velocity, or acceleration. The root-mean-square value is obtained by taking the square root of the mean square displacement during one period. For simple harmonic motion, i.e. sine wave motion, the root-mean-square (rms) value is equal to $\frac{1}{\sqrt{2}}$ times the zero-to-peak value.

Shock   Shock has no standardised definition. It is generally used to describe a process in which an impulsive force produces a sudden change of velocity or position. It usually implies some form of impact.

Simple Harmonic Motion (Sine wave motion)   Simple harmonic motion is the simplest form of oscillation. It corresponds to the motion of a piston and crank system with an infinitely long connecting rod. It consists of oscillation at one single frequency corresponding to the rate of rotation of the crank shaft.

For simple harmonic motion acceleration, velocity and displacement are related as below:

$$\text{velocity} \qquad 2\pi f \times \text{displacement}$$

$$\text{acceleration} \qquad -(2\pi f)^2 \times \text{displacement}$$

where velocity, acceleration and displacement are in comparable units and f is in Hz.

For example, for a frequency of 100 Hz the corresponding acceleration, velocity and displacement are as follows:

Acceleration $-39.4 \ m/sec^2$

Velocity $0.0628 \ m/sec$

Displacement $0.0001 \ m$

These relationships are used to obtain displacement measurements from acceleration readings and vice versa.
Note: the minus sign in front of the acceleration implies that if the displacement is upward the acceleration is downward and vice versa. The velocity is always in the same direction as the movement, hence no minus sign.

Spring Constant or Spring Rate   The force required to deflect the spring by unit distance, e.g. 1000 kg/m.

Static Deflection   The distance by which the springs or mounts under a suspended body deflect from their unconstrained position when the dead weight of the body is slowly lowered on to them.

Stiffness   Stiffness is an alternative term for spring constant.

Stiffness Control   A system is said to be stiffness controlled when it oscillates at frequencies below its natural frequency. Under these conditions it is the stiffness of the springs which controls the amplitude of vibration. Vibration isolators operated under these conditions will provide no isolation and may provide amplification.

Torsional Vibration   Torsional vibration is a twisting form of vibration where the mass effect is provided by rotating body and the spring results from the twisting of a shaft or other component. The rotating pendulum of a "perpetual" clock is one example of torsional vibration.

Transient   The short lived condition that occurs as a system changes from one equilibrium state to another in response to a change of outside condition.

Transmissability   The proportion of the disturbing force which is transmitted through vibration isolators. It is normally expressed as a percentage.

Vibration Isolator   The preferred term to describe "antivibration mounts" etc. The devices do not prevent vibration, they isolate it.

Viscous Damping   Damping in which the damping force is proportional to the velocity of movement. In theory oil or air filled dashpots provide this form of damping.

Bibliography

General

1. Broch J.T. 'Mechanical Vibration and Shock Measurements'
    Bruel & Kjaer Limited, Naerum, Denmark

2. Crede C.E. 'Shock and Vibration Concepts in Engineering Design'
    Prentice-Hall, Englewood Cliffs, New Jersey, 1965

3. Crede C.E. 'Vibration and Shock Isolation'
    John Wiley & Sons Inc. 1951

4. Crocker M.J. (ed.) 'Noise and Vibration Control in Engineering'
    Ray Herrick Laboratories, Purdue University, Indiana, 1972

5. Den Hartog J.P. 'Mechanical Vibrations'
    McGraw-Hill Book Company, New York, 1956

6. Harris C.M. and Crede C.E. 'Shock and Vibration Handbook' (3 vols)
    McGraw-Hill Book Company, 1961

7. Lazan B.J. 'Damping of Materials and Members in Structural Mechanics'
    Pergamon Press, New York, 1968

8. Morse P.M. 'Vibration and Sound'
    McGraw-Hill Book Company Inc., 1936

9. Ruzicka (ed.) 'Structural Damping'
    A.S.M.E., New York, 1959

10. Wallace R.H. 'Understanding and Measuring Vibrations'
    Wykeham Publications (London) Limited, 1970

Machine Vibrations

11. Anon. 'Sound and Vibration Control'
    Chapter 31 ASHRAE Guide & Data Book, 1967

12. Crockett J.H.A. and Hammond R.E.R. 'The Dynamic Principles of
    Machine Foundations and Ground'
    Institution of Mechanical Engineers, 1949

13. Osborne W.C. 'Attenuation of Vibration from Continuously Running Machinery'
    JIHVE, page 281, November 1960

Building Vibration

14. Allen P.W., Lindley P.B. and Payne A.R. 'Use of Rubber in Engineering'
    MacLaren & Sons Limited, London 1967

15. Anon. 'Vibrations in Buildings (Parts I and II)'
    Building Research Station Digests Numbers 117, 118, 1970

16. Crockett J.H.A. and Hammond R.E.R. 'Reduction of Ground Vibration into
    Structures'
    Institution of Civil Engineers. Structural Paper Number 18, 1947

17. Steffens R.J. 'The Assessment of Vibration Intensity and its Application to
    the Study of Building Vibrations'
    National Building Studies Report Number 19, HMSO 1952

18. Steffens R.J. 'A Bibliography on Vibration 1955-1965'
    B.R.S. Library Bibliography Number 199, August 1966

19. Steffens R.J. 'The Problem of Vibrations in Laboratories'
    Building Research Station, current paper 14/70, 1970

Journals and Handbooks

20. 'The Insulation Handbook'
    published annually by Lomax, Wilmoth & Company Limited, Rickmansworth

21. 'Applied Acoustics'
    published quarterly by Elsevier Publishing Company Limited, England

22. 'Journal of Sound and Vibration'
    published monthly by Academic Press Inc. (London) Limited

23. 'The Journal of the Acoustical Society of America'
    published monthly by the Acoustical Society of America

24. 'Acustica'
    published bi-monthly by Hirzel Verlag, Stuttgart, Germany

25. 'Technical Review'
    published quarterly by Bruel & Kjaer, Naerum, Denmark

26. 'Noise and Vibration Bulletin'
    published monthly by Multi-Science Publishing Company Limited, London.

SOUND RESEARCH LABORATORIES LIMITED
Consultants in Environmental Acoustics

Other books published by SRL include:

Noise Control in Industry
Noise Control in Mechanical Services
Practical Building Acoustics (Paperback)

Specialist Reports written by SRL engineers on the following subjects:

Noise and the Driver
Noise and Commercial Vehicles
Machine Noise Testing and Reduction
Problems of Secondary Regeneration in Air Distribution Systems
Ductborne Noise
Noise in the Woodworking Industry.

Training Courses in Noise Control run by SRL:

Controlling Noise in Industry
Controlling Noise in Mechanical Services
Practical Building Acoustics
Basic Vibration Control
Airflow and Acoustic Measurement for Building Services Engineers

2 day courses on the above subjects held at Holbrook Hall and regional centres.
Course manuals are available.

Free loan of films

Brochures on free distribution:

Company brochure - outlines consultancy work undertaken in Industrial,
Building, Vehicle Noise Control, and Laboratory fields.
Acoustic and Aerodynamic Laboratory Services
Vehicle Noise Control Services
Training Courses in Noise Control

For details and prices of any of the above information write to:

Sound Research Laboratories Limited
Holbrook Hall   Little Waldingfield
Sudbury Suffolk CO10 0TH
Tel: Lavenham (0787) 247595

Index

A

Acceleration 98
Accelerometers 92 101
Air mounts 83
Allowable floor deflections 35
Amplification 21
Amplitude 98
Articulated bellows connection 77
Attenuation through structure 18
Attenuation, Natural 18 19
Automatic frequency analysis 100

B

Beam bases 43
Bellows connectors 69
Building component 25
Building structures 26

C

Cable connectors 78
Cable 'whip' 106
Calibration (of measuring equipment) 94
Causes of vibration 11
Centre of gravity 40 44
Centre of gravity, Lowering of 33 40
Centrifugal chillers 42
Compensated bellows 76
Compound spring systems 28
Compressors 12
Concrete floors 22
Concrete inertia bases 43 44
Cooling towers 43
Cork isolators 31
Coupled modes 16 81
Critical whirling speed 12

D

Damping 4 20
Damping ratio 6
'Dead beat' damping 6
Decoupling 63

127

Disturbing force 32
Disturbing frequency 31

E

Efficiency 5
Elbow support 56
Electrical conduits 69
Equal loading 38
External vibration sources 17

F

Fan reaction pressures 39 53
Fans 41
Fans, Axial 42
Flexible collars 66
Flexible connectors 66 76
Flexible couplings 39 68 73
Flexible hose (high pressure) 69
Flexible hose installation 75
Flexible hose (low pressure) 69
Flexible hose (metallic) 78
Flexural waves 18
Floated pipe runs 67
Floor deflection 19 32
Floor loading 20
Floor spans 32
Forcing frequency 4 21 35
Frequency response curves 107

G

Ground 25

H

Harmonics 12
Height restrained coil springs 43
Horizontal translational modes 82

I

Impact and shock 13
Inertia bases 39 40
Inertia blocks 33 39 42 55 84
Internal combustion engines 40
Internal damping 19
Isolated pipe clamps 79
Isolation, Effectiveness of 5

Isolation efficiency 31 34
Isolation, Impact 85
Isolation of sensitive equipment 83
Isolation, Positive, passive, etc. 110
Isolator performance 32
Isolator selection 31

L

Linear isolators 33
Local resonance 27
Longitudinal waves 18

M

Maintained waves 18
Mass 4
Mass ratio 41
Mass controlled 4
Moments calculations 46
Mounting points 38
Mount location - equal loading 38
Mount location - five mount systems 38
Mount location - four mount systems 38 43 48
Mount location - six mount systems 38 43 49
Multi-cylinder engines 12

N

Natural dissipation 63
Natural frequency 2
Natural frequency, Of ground 25
Natural frequency - wind sway 17 25
Natural rubber 6
Neoprene isolators 31

O

Organic isolation materials 33
Out-of-balance force 4 11 13

P

Percentage isolation 21
Percentage transmission 35
Periodic motion 96
Periodic phenomena 91
Pipe connections 75
Pipe hangers 80
Pipe suspension 74

Pipe vibration 29 67
Plantroom design 63
Preamplifiers 93 99
Pre-tilted fan 54
Principles of mounting 20
Propagation 26
Pumps 40

Q

'Q' factors 20
Quick selection of mounts 43 44

R

Random signals 91
Reciprocating compressors 41
Reciprocating machinery 12
Recorders 94
Reiher-Meister curves 110
Re-radiation 21
Resilient clamping 21
Resonance 3
Resonance features 18
Resonant frequency 19 21
Resonant frequency of concrete floors 22
Resonant frequency of coupled systems 23
Resonant frequency of sensitive equipment 83
Rotating machinery 11
Rotor balance 72
Rubber expansion joint 77
Rubber-in-shear isolators 31 32

S

Servo-level controlled mounts 83
Signal analysers 93
Snubbers 33
Sound barrier mat 66
Sources of vibration 11
Static deflection 35
Steel spring isolators 31 32
Stiffness 1 4
Stiffness controlled 3
Stiffness gradient 28
Sway mode 17

T

Torsional waves 18
Transmissibility 5 31
Transmission through structure 18
Transmitted force 3
Transverse waves 18
Travelling waves 18 20

V

Velocity 98
Vertical inertia force 16
Vibration isolated pipe clamps 79
Vibration isolation of pipes and conduits 67 79
Vibration measurement 91
Vibration measurement instrumentation 92
Vibration transducers 92 99
Vibration transfer 74
Vibration, Types of 91
Vibratory motion 92
Viscous damping 6

W

Waveforms 96
Weight distribution 44 45 50 51
Wind sway frequencies 17

Y

Yawing couple 16